BLACK HOLES BUILT OUR COSMOS

by the author of
TRILLION THEORY

ED
LUKOWICH

JEPKO
PUBLISHING

Website: www.trillionist.com

JEPKO

AN IMPRINT OF JEPKO PUBLISHING, CALGARY, CANADA

Series: The first book of this series was 'The Trillionist' (sci-fi novel authored in 2013 under pen name Sagan Jeffries). 'Trillion Years Universe Theory' published in 2014 by Ed Lukowich was the initial non-fiction book outlining Trillion Theory. In 2015 'Black Holes Built Our Cosmos (Trillion Theory)' is a non-fiction book and the 3rd book in this series.

Library and Archives Canada Cataloguing in Publication
Lukowich, Ed – Black Holes Built Our Cosmos.
ISBN: 978-0-9918408-4-7 (e-Book ISBN: 978-0-9918408-5-4).
Cosmology. FIRST EDITION. Printed by CreateSpace.

Black Holes Built Our Cosmos

TABLE OF CONTENTS

PREFACE

'Astronomers admit that they are re-thinking our universe. Large cracks are appearing in Big Bang.'

"**Back in 1998,** I'd been intensely studying and re-theorizing our universe for well over a decade. Yet, my newly formulated *Trillion Theory*, depicting our universe as some ancient trillion-year-old growing entity, found itself collecting dust – unpublished.

As a busy Olympic coach, icy time slid by. Then finally, 15 years later in 2013, I took my first baby step with the release of some of my new ideas through my sci-fi novel *The Trillionist* (published under a pen name Sagan Jeffries). The novel told the story of an unscrupulous entity, far older than dirt, bending and cheating normal planetary life.

Readers of the novel seemed to like my far-out ideas, so in 2014 my 'Trillion Years Universe Theory' was released. Thereafter in 2015, 'Black Holes Built Our Cosmos (Trillion Theory)' delved deeper inside of Trillion Theory showing the major role played by black holes during the trillion-years of our cosmos from small beginnings to present humongous size.

Trillion Theory solves mysteries which continue to stump Big Bang. Both Bang and Nebular incorrectly estimate our universe as only 13.7 billion years old.

My *Black Holes Built Our Cosmos (Trillion Theory)* uncovers new controversial and fascinating theory as to how matter first formed in our universe and then how our own planet, our solar system, and billions of galaxies came to be."

**B
L
A
C
K

H
O
L
E
S**

Introduction to the author Ed Lukowich

Foreword by Brian Hades of EDGE Science Fiction and Fantasy Publishing. Brian published Ed's futuristic sci-fi novel *The Trillionist* under pen name Sagan Jeffries.

Brian says: "When Ed first approached me to publish his science fiction novel and support his ideas about our universe, I do admit to being somewhat skeptical. But, while rookie Ed brandishes no significant astrophysics pedigree, he has a most unique longtime developed perspective of our universe. His new model depicting how our cosmos came to be is indeed most intriguing. The more I saw of his new theories when he wrote his first theory book *Trillion Years Universe Theory,* the more I became enthralled by his Trillion Theory as to how our universe began and grew.

As a self-proclaimed theorist, Ed takes our universe way past any Big Bang of 13.7 billion - to a trillion years – where he demonstrates how miraculous scientific inventions were behind the mechanisms which built our cosmos. Black holes play a most major role in his *Black Holes Built Our Cosmos* as the organizational builders of galaxies and solar systems and the stars, planets and moons within those systems. I recommend his books as incredible eye-opening reads."

Now, a word from the author Ed Lukowich:

When I first began to formulate *Trillion Theory back* in 1998, the only license I wanted was that of being human, for everyone has the right to search for life's answers. My first task was to attempt to figure out a new theory to explain our physical universe, as the widely accepted, yet haphazard Big Bang theory did little to quench my thirst.

Over the past 15 years, each puzzle that an astronomer or astrophysicist ran into seemed to always do more to confirm the validity of my Trillion Theory. Here are a few examples:

In 2004, a Gemini Telescope program found that galaxies seemed more fully formed and mature than one might expect to find inside a supposed young 13.7 billion year universe. A quote from Dr. Robert Abraham, Department of Astronomy and Astrophysics, University of Toronto, "We are seeing that a large fraction of the stars in the universe are already in place when the universe was quite young, which should not be the case. This glimpse back in time shows pretty clearly that we need to re-think what happened."

Dr. Patrick McCarthy, Observatories Carnegie Institution, added, "It is unclear if we need to tweak the existing models or develop a new one in order to understand this finding."

Then in 2014, astronomers found in our Milky Way Galaxy a star which they named Methuselah. At 16 billion years of age it was older than the supposed 13.7 billion year origin of our cosmos. Big Bang had no way to explain Methuselah.

As a newly presented model, Trillion Theory shows why older stars exist in a supposed young part of our universe. Trillion Theory displays how various solar systems formed at different rates over the past trillion years. Thus, in Trillion Theory, older stars can exist alongside young stars, just like in a recycling growing forest where an old tree can stand tall beside a small young sapling. Therein, Trillion Theory takes recycling to the stars, to solar systems, and to galaxies.

As a brand new theorist, in *Black Holes Built Our Cosmos* I take us on a journey which goes back a trillion years; to an ancient past where we see firsthand the small beginnings of our cosmos, when the first planet, star, and solar system had their origin. Then, we follow the recycling growth of our cosmos to the present. Black holes played a major role.

Right, wrong, or just ideas to expand one's thinking:

As the reader, your first thoughts when you read my new universe theory likely are, 'What cockamamie stuff is this? As absurd a theory as I've ever heard of. It's not plausible that our universe began as Trillion Theory proposes. Too crazy.'

As the founder, I readily admit that *Trillion Theory* is very much a maybe theory at this stage. On the other hand, I've a deep belief in my *Trillion Theory.* Yet, I'm a very open-minded individual, with a love of life, great sports and good humor. The other day, a friend told me that my theory had about as much chance as his flying and landing on Pluto. To which I decided to rename my dog Pluto, just to up his odds.

However, as a sportsman with a wacky sense of humor, once in a while my serious side (*Trillion Theory*) shows up. Therein, let's take a serious look at Trillion's possibilities.

Right: It'll take a prolific discovery to prove Trillion Theory right, such as an announcement from an astronomer, 'Yes, we witnessed that a black hole can grow a sphere such as a planet around itself.' Such a discovery would mean that my years of formulating Trillion Theory would have been fruitful.

Wrong: What if Trillion Theory is dead wrong? Then, I'd be the first to welcome a better cosmic explanation. So, for the present my *Universe Trillion Project* is a go (it is my pursuit to discover what we're doing on this planet in this cosmos – maybe it's just a nice holiday to either enjoy or hate).

Partially right: What if Trillion Theory has a few new things right, and some wrong? That would be a victory! For, even a new 'something right' could have huge implications towards solving the mysteries of our physical universe.

Food for thought: What if no one can prove Trillion Theory at this time, or its ideas are too new for acceptance? Then, perhaps Trillion Theory is merely food for thought to help expand thinking along a road to understanding our cosmos.

CHAPTER 1
PREVIEW: WHAT IS
TRILLION THEORY

'A brand new theory vastly expands the boundaries of our minds. And, no great idea was ever immediately accepted - most were initially thought of as ridiculous.'

Trillion Theory (*Trillion Years Universe Theory – TYUT*) is a new universe theory proposed by new self-proclaimed theorist Ed Lukowich. In Trillion Theory, the author proposes a totally new and very different explanation of our physical universe, and this new theory refutes the Big Bang.

Black Holes Built Our Cosmos further explains, within the confines of Trillion Theory, the specific role which black holes have played with regard to the origin, age, growth, recycling, organization, and operation of our cosmos.

Now, the prime purpose of this chapter is to provide you the reader with a tasty preview of what to expect. I start by saying that Trillion Theory is detailed, as it should, for our universe is complex beyond belief to the extent that the normal 13 seconds it takes to explain theories such as Big Bang or Nebular simply won't suffice. So, I implore you to read the entire book before formulating your final opinions.

In preparing this preview, I sought brevity coupled with a broad look at the role which black holes play in Trillion Theory. How much is enough, or too much in a preview?

This reminded me of an old story. 'It was the worst blizzard seen in years. Yet, the dedicated Farmer braved the cold as his horse-drawn sled made it to the church on time for the Sunday service. One look inside told the Farmer that he was the sole parishioner in attendance. As the Minister approached, the Farmer asked, "Will there be a service?" The Minister relied, "God told me long ago that if a farmer gets only one cow to the lot, he still feeds it." So, the Minister did the full service, A to Z - while, the Farmer endured. At the conclusion, the Minister asked, "How was my service today?" To which the Farmer replied, "If only one cow comes to eat, I still feed it, but I don't throw off the whole load."

This Preview begins with enough fuel to build a fire.

Trillion Theory proclaims our cosmos to be a trillion years old, thereby far older than the 13.7 billion years as estimated by Big Bang. For, 13.7 billion years is just the age of the current rendition of stars in our sky.

Trillion Theory is far different than Big Bang. So, before delving into the vital role which black holes play in Trillion Theory, it is necessary to look at the 'big picture' of our universe to see how Trillion Theory depicts differently.

In presenting Trillion Theory, this chapter provides an entry-level 'big picture' preview. (Later chapters give greater detailed descriptions and explanations).

To begin, we move to a unique perspective vantage point, outside of our known universe. From there we get a *big picture* look, and we can also *zoom inwards* to see the specifics comprising the inside. From these various vantage points we can review some things which we already know. Along the way, Trillion Theory can add-in new cosmic ideas.

Shape of the entirety of our universe.

According to Trillion Theory, the general shape of the entirety of the 'space' of our universe is not round but rather oblong, longer in one direction; similar to the shape of an egg or even more like the shape of a cucumber. The reason for this elongated shape will be explained in this book.

The entirety of our universe's space is a cucumber shape.

From outside of this entirety of space we are able to see inside to where billions of galaxies predominantly exist.

In the space of our cosmos there exist billions of galaxies. According to Trillion Theory, some galaxies are hundreds of billions of years old; some are younger or just forming; while some others are reforming.

According to Trillion Theory, the general location of most of the billions of galaxies within the entirety of the space of our universe is generally away from the center; somewhat closer towards the outer perimeter edges of space. The reason for this will be explained in this book.

Of course solar systems exist within these billions of galaxies and of course we all know that our own particular solar system is in the Milky Way Galaxy.

A rough model of our present universe, with billions of white-colored galaxies positioned outwards away from the center.

Note: Such an elongated oval shape for the entirety of the contents of our universe indicates that the Big Bang never happened as a bang explosion would have produced a roundish spherical isotropic shape.

Outer boundaries of our universe.

According to Trillion Theory, the outer limit of space is not the outer limits of the entirety of our physical universe.

Astronomers consider that the outer limit boundaries of our universe are as far as the most distant galaxies or as far as 'space' extends outwards. Therein, our physical universe is supposedly everything inside of this space area. A key relevance here is that if this were correct, then all the available gases to coalesce and form spheres in galaxies is already inside of this space area. That's all there is. And if this were so, the absolute amount of matter in our universe would be constant with no means to increase.

On the contrary, according to Trillion Theory, out past the edge of space there exists a supply of material readily available to be deployed to continually increase the matter content inside the space of our universe, and furthermore to actually increase the size of the entirety of space.

According to Trillion Theory, utilization of this external supply of material has been going on for the past trillion year history of our physical universe.

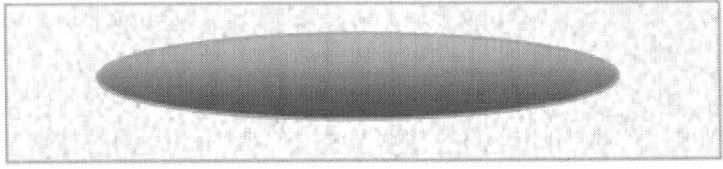

Out past the outer limits of space exists a supply of material (shown in white) ready to be used to continually increase the count of solar systems and galaxies inside our cosmos.
(And as explained later, to also increase space's size).

What is outside of this entirety of space?

According to Trillion Theory, more energy is still available from outside the outer boundaries of space to be brought into and used to build more spheres, solar systems, and galaxies in our universe. The importance of this new idea is that our universe is able to continually grow in size, having more available energy supply from without to build more and more galaxies inside of our universe.

Therein, Trillion Theory extends the outer boundaries of our universe showing how the limits of space are continually expanding outwards. You ask, outwards into what?

Depicted below, Trillion Theory shows the existence of a gigantic sea of light energy surrounding the space of our universe. This sea or ocean of static (frozen) light energy is continually available for the building of new spheres, solar systems, and galaxies to forever grow our universe.

A decent analogy to help us understand this concept would be a normal mining operation. Miners burrow deep into the outer walls of a mine shaft, reaping ore which is used to build things, while at the same time ever-extending the overall space of the mine.

According to Trillion Theory, as more and more energy is broken away and brought inwards from the outer static ocean, more and more solar systems and galaxies are built. As well, as this static ocean of energy is dug into, nothing is left where that energy was depleted, nothing that is except for the newly created nothingness of space. Thus at day's end, it is possible to build more galaxies and simultaneously to extend space further afield outwards. This outward thrust means that the overall movement of objects in space is outwards towards the outer perimeters of space.

Later, Trillion Theory further details this entire process.

Zoom inwards to see the contents of our universe.

Inside of our physical universe, space comprises over 99% of the area. Thus, huge empty distances of black space, void of matter, exist between galaxies. However, although empty, this space does provide a means of transportation for galaxies to move around and for light to travel through.

Why such an enormity of space in our physical universe? Later, Trillion Theory answers this pertinent question.

Also inside of our physical universe, most of the 1% of matter content making up our physical universe is housed in oasis-type islands in space called galaxies. Trillion Theory depicts these galaxies as islands since generally it takes even fast-moving light long travel times between galaxies.

According to Trillion Theory, some of these galaxies are hundreds of billions of years old, while some are younger, and some are just forming or reforming.

The most common galaxy is the spiral galaxy. There are billions of these spiral galaxies, all isolated islands in motion, spinning either clockwise or counterclockwise around the supermassive black hole located at the central bulge of the galaxy. Later in this book, Trillion Theory will answer why there are these two differing directions of spin.

Zoom further inwards to view one galaxy.

A spiral galaxy is pancake or frisbee shape with a bulge at the center. Trillion Theory will show how the laws (rules) which govern our physical universe provide this flat shape.

At the hub of the spiral galaxy there is a supermassive black hole, using its powerful spin and its gravitational pull to hold the entire contents of the galaxy in a formation; and gigantic stars forming the body are held close in around the supermassive black hole; while whirling around the outer part of the galaxy are 5-6 spiral arms. The body and arms all revolve in the same direction as regulated by the supermassive black hole organizer. Trillion Theory will show how cosmic *laws* command this spiral formation.

This spiral galaxy houses millions of stars with solar systems (planets and moons) in orbit around themselves. Each star orbits the same direction as that of the rotational spin of the supermassive black hole organizer of this galaxy.

Thus, these solar systems are in constant motion, each orbiting either clockwise or counterclockwise around a dictating central sun. The revolve direction, either clockwise or counterclockwise for the entire solar system, is dictated by the direction of spin of the sun located at the center of the solar system. Note: Basically, if any other current theory other than Trillion Theory were true, then all galaxies and all solar systems should have only one direction of spin, either all clockwise or all counterclockwise. But, Trillion Theory shows 'direction of spin' is determined by *Cosmic Laws* set out at organizational-time of any particular solar system.

Now as we know with our own solar system, all the planets of our solar system revolve in orbit in a similar direction around a central sun. The sun of our solar system spins counterclockwise on its own axis and with its massive gravitational force holds 8 planets in a counterclockwise direction in orbit. Each of these 8 planets can hold a moon(s) in orbit. While Earth has only one moon, larger planets have many moons, with over 100 moons in our solar system.

However, each of the individual planets (and moons) within a solar system can vary in size, composition, and also can set their own direction of rotation (pivot spin) on their own axis. This feature is witnessed in our solar system where

6 of the 8 planets rotate on their axis in a counterclockwise direction while the other 2 spin clockwise on their axis. Later, Trillion Theory will show how *Cosmic Laws* govern this varying direction of rotational axial spin for individual suns, planets, and moons. Note: If Big Bang or Nebular Theory were true, then all the individual planets and moons within a solar system should follow only one axial spin direction.

Zoom inwards even further.

Zooming from macro into the micro world on all of the planets, moons, suns, stars, solar systems, and galaxies comprising the matter of our universe, we find that all the matter forming them is a complex subatomic world of continuously spinning atoms. Over 100 varying types of atoms, each with its own definitive subatomic structure, have been discovered. Many of these building block atoms combine into more complex molecules providing a vast array of matter. Neither Big Bang nor Nebular Theory can explain why atoms exist – while Trillion Theory can explain.

Now, zoom partway back out to see what we have. A good analogy for galaxies and solar systems is a hotel.

Imagine the galaxies of our cosmos as island hotels. The galaxy is a rather permanent hotel up to hundreds of billions of years old. Occupying the rooms of this ancient galaxy hotel are non-permanent solar systems which have a shorter lifespan (normally an upper limit of 15 billion years).

Every unit solar system is independent of every other system in that galaxy. Like with a hotel, any one particular solar system can check out of its room when its own sun goes Supernova, without affecting any other solar system.

When a sun ages out, goes Supernova and explodes near the end of its 15 billion years, and that solar system is destroyed, it makes room for a new tenant in that particular room of the galaxy hotel. However, Trillion Theory states that deceased solar system isn't totally dead, as it recycles into the advent of a new larger solar system (growth in numbers via black holes), or spits into two new solar systems making for two rooms replacing the one old room in that quadrant of the galaxy. Either way, it replenishes that vacant room, increasing the solar system population of the galaxy.

And, every solar system ages-out differently, dies, loses its contents, and then prepares to recycle into a new solar system which will occupy that area in the galaxy for the next approximate 15 billion years.

The absolute stalwart throughout this recycle of the many solar systems within a galaxy is the old age supermassive black hole at the center of the galaxy. Over hundreds of billions of years, the old supermassive at the center of the island galaxy survives any and all supernova within the many solar systems living inside of its island borders.

True age of various components of our universe.

According to Trillion Theory, our universe began at least one trillion years ago. Trillion Theory also further depicts varying ages for various components of our universe. While Trillion Theory calculates the entirety of our cosmos to be one trillion years, various galaxies within the universe are anywhere from 200 billion to 800 billion years old. Inside these galaxies, various solar systems and their resident spheres (stars, planets, moons), are anywhere from new to 15 billion years old. Usually, solar systems have normal top limits to their cycles approximating 15 billion years.

The 15 billion normal recycle age of solar systems aides Trillion Theory in estimating the overall age of our physical universe. Doing the math, 67 cycles x 15 billion years per cycle, equals to a trillion year universe history. Fifteen billion years x 67 cycles each on a doubling progression from 1 star gives 73 quintillion stars in our present universe.

Structurally, the Hubble Space Telescope shows that each of these humongous galaxies has a supermassive black hole at its bulging central hub. During the extremely long epoch of universe history, billions of gigantic galaxies, far separated from one another, formed up into islands, acting like 'hotels.' Within each galaxy, there formed multitudes of quite-large units known as *solar systems.* - acting like rooms in the island hotel.

Where black holes enter and fit into Trillion Theory.

Trillion Theory proclaims that **black holes** are prevalent in a vast array of places in all the galaxies of our cosmos. These black holes come in a vast variety of sizes:

- Trillion Theory sees a **supermassive black hole at the hub of every galaxy.** These massive bulging supermassive black holes possess the mightiest gravitational forces, capable of holding large stars near to themselves and millions more stars and solar systems in the spiral arms of the galaxy.

- Trillion Theory proclaims that hidden a cloaked **extra-large black hole is at the core of every star.** These stars (suns) are at the center of all the solar systems of a galaxy. These extra-large black holes hold all the planets (and moons) of that solar system in continual revolutionary orbits encircling the sun which is central to the solar system. Note: Nearly every star utilizes its XL black hole spinning gravity to form a solar system of planets around itself.

- Trillion Theory proclaims that a cloaked **black hole (either of large, medium or small size) is at the core of every planet and moon in a solar system.** These **lesser sized**

black holes vary in size ranging from a **large black hole** at the core of a large planet down to the **smallest black hole** of a small moon. However, any planet acting on a proximity basis near to itself, can trump the gravity of the sun of that solar system. The larger a planet, the more moons it can have under its control. Our Earth, with a **medium-to-small size black hole** in its core, holds only one moon in orbit. Whereas, larger Jupiter with a **large black hole** in its core is capable of holding 63 moons in orbit.

▪ Therein, Trillion Theory claims: *'there exists a supermassive black hole at the hub of every galaxy, and smaller black holes at the core of each star/planet/moon in our cosmos.'*

Again, Big Bangers are still trying hard to fit the discovery of black holes into Big Bang thinking. But, black holes simply don't fit into Big Bang theory as there is no place for them.

Whereas, Trillion Theory shows that our universe *could not have come to be without black holes.* Black holes are absolutely necessary to our universe. So we ask, "What then are black holes doing in our universe? Why do they exist?"

My new Trillion Theory shows how black holes operate in a machine-like manner, building and at the same time bringing orderly organization to galaxies, and solar systems. Black holes have been doing their job for a trillion years, recycling the matter and energy of our universe.

Black holes have built billions of solar systems in the past, and our solar system is simply one of the latest current.

Unfortunately, all the past solar systems of past ages have left no trails to sniff, or graveyard head stones to mark, proving that they ever existed. Yet, Trillion Theory shows that they did exist. As in the ancient past with our universe, today multitudes of black holes are busy building, recycling, organizing and bringing order to our cosmos.

The larger the sphere, the larger the black hole at its core; and on the lesser side, the smaller the sphere the tinier the black hole at its core. Approaching this from another angle, the larger the initiating black hole the greater is its gravity and the larger the resultant body it can build around itself.

To more clearly differentiate the size progression of black holes: a supermassive black hole is at the hub of a galaxy; all the lesser black holes (XL, large, medium, small, and XS) are at the centers of the stars (suns), planets and moons.

Supermassive black holes are formidable compared to all the other lesser sizes of black holes. And, supermassives have far greater prominent defined roles compared to all the lesser black holes in our cosmos.

Supermassive black holes stand out on their own, totally visible. The reason they are totally visible is that they function differently than any lesser black hole of smaller size. All lesser black holes are either cloaked (residing inside of spheres which they have built around themselves), or sometimes astronomers will spot them as naked black holes which have shed their contents and become visible since they are no longer hidden away inside of a sphere. However, this naked feature may only be temporary till they fill back up and re-cloaks itself in the recycling process.

Note: You might ask: Why don't the black holes within the spheres of a solar system devour and unravel their sun?

Trillion Theory answer: The black holes at the center of all the spheres in a solar system are cloaked and in a *holding of matter* mode rather than spinning energy into matter mode. Basically all the bellies of those current lesser black holes are plumb full with matter, unable to eat more, whereas naked empty black holes are eaters and destroyers of nearby suns.

Trillion Theory presented here shows:

• How our universe began a trillion years ago.

• Our cosmos is far older than Big Bang's 13.7 billion years.

• How moons/planets/stars were born.

• How spheres recycled approximately every 15 billion years.

• How our recycling universe doubled in number of spheres with each cycle to its present size.

• How our recycling cosmos is of infinite design.

• How all the present moons, planets, stars are in different phases in their individual cycles.

• Why our entire universe shows itself to be more linear in shape than spherical.

• How galaxies receding away from each other is an outward pulling force (not from an internal explosion).

Foundations of new Trillion Theories:

• Trillion Theory places LIGHT as the material at the forefront of our universe. Ever humble, light displays more of its many talents, upping its phenomenal properties. Trillion Theory will demonstrate how light is incredibly spun into matter.

• Trillion Theory drastically changes the duty role of BLACK HOLES, showing them to be unique engines which built the spheres of our universe by spinning light into matter. They are catalytic engines of the recycling processes of the spheres in our universe over the past trillion years. They give credence to the evolutionary growth of our universe.

• Trillion Theory provides a better explanation of what SPACE is and how it formed; and, how TIME is a specialty item designed specifically for our universe.

Where gravity enters the picture.

When we talk black holes, we must talk about the gravity they produce. For, Trillion Theory proclaims that black holes are the dominant players providing gravity as we know it.

Throughout, gravity is the great constant associated with each and every black hole regardless of size. All black holes pivot and spin furiously on their personal axis, and all black holes from this spin have a definite method of generating gravity which makes them into swallowing pits for any energy passing nearby. This gravity also gives naked black holes an absolute pulling force capable of even unraveling nearby stars, sending them early into Supernova action.

The importance of various levels of gravity?

The size of black holes, and their gravitational net which surrounds them, are prominent factors in determining where the various black holes position during and after the battle within the building of a solar system.

Trillion Theories show that at certain times in areas within our cosmos, there can take place either a 'gravity battlefield,' or a 'gravity cooperative,' depending upon the stage in the development of a solar system.

A full out battle for energy to spin into matter takes place between the existing naked black holes present at the time when the structural arrangement of a solar system initiates.

Thereafter, once a solar system is finally formed, the black holes residing at the cores of all the spheres according to set *Universe Laws* in a cooperative manner via their gravities so that the solar system can function smoothly. The extra-large black hole within the sun of the solar system controls the entire system around itself, while the arrangement of planets and moons can vary with each different solar system. Hence, solar systems are akin to snowflakes, where no two solar systems are ever exactly alike.

The solar system we live in is presently in a cooperative mode, with each sphere in our solar system having a full belly and full body around the black hole at each core - there is peace in our solar system.

However, Hubble Space Telescope can find multitudes of distant Supernovae, and each Supernova is in its explosion stage wiping out the planets and moons of its solar system.

It is within the grave-like remnants of this old destroyed solar system where the *true story of Trillion Theory is found*, as the surviving naked black holes which had been at the core of every sphere now commence their battle to spin new energy into matter around themselves. The competition to build a new larger solar system has begun on that spot. The finding of this battlefield would be proof of Trillion Theory.

Survival of the fittest fiercely takes place in the battlefield between the naked black holes as they use the power of their spin to attempt to gain a major position within the new solar system which they are building. The largest extra-large black hole of the entire group of naked black holes has the best opportunity to someday become the central sun.

Supermassive black holes, as the central hub of galaxies, have the greatest gravity.

Each supermassive black hole features a tremendous rate of spin on its axis thereby creating a powerful gravitational force around itself. This force is able to hold a multitude of stars (and their solar systems) close in around itself and also in the galaxies spiraling arms.

Gravity provided by the supermassive powerful black hole at the central hub of the galaxy keeps the entire galaxy (hotel) together and spinning as one.

Next, some possible support for Trillion Theory.

Astronomers have been able to identify and observe millions of supermassive back holes at the center of galaxies. Supermassive black holes are indeed the most incredible powerful feature of our universe. Astronomers are learning more each day about how these supermassives operate and then bring order to a spiral galaxy - basically, doing their designated organizational job.

History associated with Trillion Theory shows what occurs within a spiral galaxy over hundreds of billions of years. The main operative of the galaxy is the supermassive black hole at the hub. In its master's role, the supermassive is designed to endure and outlast many recycles of the contents of the galaxy, and it can be well over a half-trillion years old.

Whereas, the units of the galaxy, namely all the solar systems existing around suns have shorter life spans. Solar systems recycle approximately every 15 billion years because the life span of any particular sun (star) at the center of the solar system can vary before it dies going Supernova. With each Supernova, a unit solar system is destroyed and a new solar system begins to grow to replace the old solar system.

Supernovae ultimately produce an increased number of planets and moons with the recycle into a new solar system, or a split into two solar systems. Thus, growing the galaxy.

This author states that the happenchance associated with a Big Bang, followed by a Nebular effect, could not have built these prominent island galaxies. Also, Big Bang can't account for the supermassive black holes at galaxy hubs.

Whereas, Trillion Theory credits black holes as being the official builders of galaxies and solar systems. Trillion Theory adds that black holes are the universal suppliers of gravity.

Gravity, it's such an exciting subject, and such a vital part of black holes within Trillion Theory. It is gravity which holds us on Planet Earth so we never fall off. Gravity acts upon our bodies 24/7, day and night, everyday. Find any star, planet, moon, solar system, or galaxy, and gravity is always present.

But, what really is gravity? Where does it come from? Why is it so important to our universe?

For Trillion Theory to properly unriddle gravity, let's examine just how gravity stacks up in various locations in our cosmos. To do this, we go inside of a gigantic spiral galaxy, and then systematically follow the progression of gravitational force's outwards starting from the hub:

- The absolute greatest gravitational force observed so far in our universe is that of a supermassive black hole at the central hub of a galaxy. These black holes are the largest in our universe. As the bulging swirling hub of a galaxy, a supermassive can hold many large stars on its near perimeter, and millions more in its distant spiraling arms.

- The massive stars surrounding a supermassive black hole possess the second highest gravity. These massive stars assist to hold the entire galaxy together.

- Next strongest level of gravity is within the starry suns which are at the center of all the solar systems of the spiral galaxy. These suns prevail throughout the galaxies body and also the encircling spiral arms. Each sun of a particular solar system holds planets/moons in orbits encircling the central sun of that solar system.

- Final lesser levels of gravity are owned by the planets and moons of a solar system. These spheres are subservient to their sun. However, any planet can use its own gravity to keep nearby moons in orbit, thereby trumping (on a proximity basis) the sun's gravity. The larger a planet, the more moons it can have under its control. Earth, (a small-medium size planet) holds one moon in orbit. Whereas larger Jupiter (greater mass and gravity) holds 63 moons.

What should the reader expect in this book? Is Trillion Theory difficult to understand?"

No, not at all difficult, no crazy formulas to digest. Only formula is Einstein's $E=mc^2$. So, anyone can read new Trillion Theory and grasp the concepts, even if the reader agrees or totally disagrees with any of the new concepts.

For example: In our lives we see things in a certain way. For us, when something is made, it has to be put into something. With our universe, it's easy for us to presume that the galaxies and solar systems of our universe came to be in space. That space was there ready and waiting.

However, Trillion Theory sees space differently. And at the outset you might have a difficult time giving any credence to Trillion Theory's concept that space was not there ahead, but rather is a leftover byproduct, namely the emptiness left behind when matter formed into the spheres of our universe. That all the weight, heat, light, movement, gravity, basically everything, is spun into and onto the spheres of our universe and space is the empty byproduct left behind.

As of this precise moment, you likely say, "No way, that's not possible. Space had to be there ahead of anything else."

I say, "Trillion Theory will redefine space."

Now, the **scope** of Trillion Theory is wide dealing with all aspects of our cosmos: the age, shape, size and perimeter of our universe; early days of our universe; long history of trillion year old universe; how galaxies, solar systems, stars, planets and moons formed; the universe laws pertaining to the orderly organization in solar systems and galaxies; how atoms formed; what space is; and how time was imported.

Trillion Theory will provide greater detail as to how black holes have built up our universe over the past trillion year history. Inside of Trillion Theory, there are several theories which work to comprise the entire theory.

Here is what you can expect to find in this book.
1. True purpose of black holes in our universe.
2. Trillion Theory showing how black holes built our universe's galaxies and solar systems.
3. A look at the outside around a black hole.
4. A look inside of a black hole (never seen before).
5. The Laws by which black holes operate.
6. Possible future proofs for 'Trillion Theory.'

Our eyes can fool us.

From the read so far, you have likely gathered that Trillion Theory has many parts to its entire theory. Thus far you most probably have far more questions and doubts - than buy-ins. That's only natural when reading a new theory, for example:

Centuries ago, dating further back in Earth history past 1500 A.D. people of this planet were really fooled by their eyes. They stood on ground and watched the moon and stars move across the sky at night. It was only natural for their eyes to tell them that they were the constant non-moving center and that the objects in the sky were the movers. Wrongly, their eyes convinced them that they were the center of a small universe which revolved around them. Later, Copernicus and Newton proved that Earth actually spins making our sky appear to move and that the sun is the real center of our solar system, Earth being but one planet.

Again, people's eyes fooled them into believing that our Earth was flat. There was fear of sailing ships falling over the edge. Of course we know today that our Earth is roundish.

Today, are our eyes fooling us again, even with the great benefits provided by our technology and even with the advantage of a Hubble Space Telescope?

As a test, take a look at everything on planet Earth and state whether it lives, dies, and recycles indicating that it is organic, or whether it is a lifeless inanimate inorganic object.

My subject here is recycling and how organics grow, die, and recycle (example: an organic such as a tree is part of a recycling process into another tree or more trees). But on the other hand, an inorganic such as a rock or a boulder doesn't have the means to grow, die, or reproduce, so we believe that it can't be part of any recycling process. In a similar manner our eyes tell us that planets and moons are like rocks, with no means for a recycling process.

But, Trillion Theory states that all physical matter within our universe (regardless of organic or inorganic) has a capacity to be part of the recycling process active within our universe. Some recycling, such as an animal giving birth, occurs in a very short time frame and is easily visible and recognizable to human eyes. Whereas, with an inorganic such as a rock, planet or moon, recycling may not be visible for billions of years to come. Yet, someday our sun will go Supernova and destroy our solar system, and all of the inorganic contents within our solar system will revert back to energy available to recycle into brand new matter. The naked black holes which survive the Supernova of our solar system (one at the core of every sphere of our solar system) will battle against one another to spin that released energy back into new matter to form the next solar system.

So, a vital part of Trillion Theory is that surviving black holes have the capacity to be the recyclers of our cosmos, and they also have the built-in feature of replication. They are alive in a most unusual way.

That phrase, 'alive in an unusual different way' may seem strange at the outset, but, think about life on Earth for instance and note the differences in 'alive.' A human is a different alive than a plant, an insect, a virus, or a bacterium. Also, there are differing means of reproduction.

So, we need to expand our thinking from Earth and apply many of those same beliefs to our entire cosmos. If life, reproduction and recycling are possible on Planet Earth, why shouldn't they be possible across our entire universe, even in new ways we have never thought of before?

Trillion Theory attacks Big Bang.

Trillion Theory presently takes what is reportedly seen and discovered through powerful telescopes and provides new and differing interpretations than those provided by astronomers who man those instruments. Generally, most astronomers attempt to have everything fit neatly into the widely accepted theory known as Big Bang.

Instead, Trillion Theory says 'no' to Big Bang and puts forth a new theory stating that our universe is much older than 13.7 billion years (13.7 being the age of the oldest stars in the current rendition of our cosmos). With Trillion Theory, the star content of our universe has grown over the history of our universe, having increased in numbers with each recycle of the stars. This recycle occurs in multitudinous parts of our cosmos at various paces (15 billion years being the approximate average life span for an average solar system). Trillion Theory deals with the intricacies of an evolving-growing universe where black holes are both the machine-like builders and also the recyclers of stars, planets, moons, and solar systems within galaxies.

This preview to Trillion Theory acts as a glimpse into the scope of this book and hopefully will help the reader to determine a course of action in reading further.

If you already possess a keen interest in our universe, you might find the new 'out-of-the box' ideas compelling and feasible. Or, you might trash them in an instant as they go against most universe ideas which Big Bang has the majority of people believing as gospel.

My point is that Big Bang and Nebular, the leading theories at present, have many things wrong. And, Big Bang has a stranglehold and a powerful influence over all other backup related theories. Unfortunately, 85 years ago, when an astronomer noticed galaxies moving apart, the explosion based Big Bang slowly moved into prominence and has stayed with us for over 65 years up to the present.

History shows how the Big Bang became a theory.

Back in 1929, Edwin Hubble (Cosmic Expansion) observed that galaxies were indeed redshifting on the color spectrum, indicating galaxies were receding away from each other in a definite expansion of our universe. Then in 1931, Georges Lemaître proposed Hubble's redshift findings were due to an explosive origin of our universe. He postulated that a central singularity of matter exploded dense hot contents outward at the origin of our cosmos. Hot eddies of gas swirled, cooled, and contracted over time to form spheres.

In 1949, English astronomer Fred Hoyle, who never liked Lemaître's theory, in advertently in a BBC radio broadcast named the theory. Hoyle rejected the theory as 'cartoon physics' and mockingly coined the phrase *Big Bang* when referring to the explosion theory as a ridiculous hoax.

Then, the real clincher for Big Bang came in 1965. Arno Penzies and Robert Wilson showed that the glow from 'cosmic microwave radiation background' was almost exactly uniform in all directions. This supposed aftermath of an explosion, began very hot leaving behind a microwave glow and gravitational waves indicating expansion. With this, Big Bang vaulted into being the accepted origin theory. Since then, new discoveries must fit into Big Bang, even if they require twisting. Or, they are shelved.

This author sets out to disprove the Big Bang

This author contends that Big Bang is wrong, and that there are different explanations for 'redshift' and 'cosmic microwave radiation background.' I will show how 'isotropic' (namely the same in all directions from a center which should occur from a Big Bang explosion), is not the true nature of our cosmos. And today, most astronomers are agreeing by predicting that the shape of our entire universe is more linear than spherical.

Neither Big Bang nor Nebular can properly account for the linear shape seen in galaxies and solar systems, where spheres are arranged in a flat configuration extending nearly straight outwards from the center.

I contend that astronomers haven't looked hard enough to attempt to find another more plausible reason other than Big Bang for this outward expansion of our universe and for galaxies receding away from one another.

For instance, Big Bang (the explosion advocate) bitterly fails to explain spin. So, it recruited Nebular Theory (the advocate for gas-cloud nebulae which supposedly whirled and then coalesced spinning into the spheres, solar systems, and galaxies of our universe). However, both Big Bang and Nebula are far off from being believable.

As an independent theory writer, I'm not constrained by Big Bang. Trillion Theory goes back to see the aged history of the recycling of stars over the past trillion years.

Also, Bang and Nebular can't explain the spin witnessed

in the macroworld of galaxies, solar systems, stars, suns, planets, moons, and even down to the microworld atoms. Trillion Theory can, comprehensively showing how powerful black holes are the machine-like power engines responsible for building, growing, and organizing, and spinning the contents of our universe over the past trillion years.

Comparing Big Bang to my Trillion Theory.

Unfortunately, Trillion Theory takes longer to explain than a short Big Bang or Nebular Theory - for good reason. Our universe is complex and tricky beyond belief, so it takes more than the 13 seconds Big Bang takes to explain itself.

First, here goes the Big Bang theory explanation of our universe: 'Hot matter supposedly exploded from a central source across space, where the coldness cooled the hot matter into the many spheres of our universe.'

But, there were and are a zillion unanswered follow up questions. When Big Bang failed to answer those pertinent questions, Nebular Theory was recruited to support the Big Bang as follows: 'After the Big Bang explosion, the hot interstellar gases spun in space like a Nebula, and as the matter spun it coalesced into the hardness of planets and moons (smaller and cooler) and into hotter larger spheres called stars. Since stars were larger, they supposed didn't cool in the coldness of space. Then, the gravitational spin of the nebula brought the smaller spheres into orbit around the larger suns forming solar systems and galaxies.'

Once again, zillions of unanswered questions result from inadequacies of Big Bang and Nebular Theories. Basically, humans had been provided with a no-brainer (easy to accept) universe theory via Big Bang. That theory sufficed over the past 65 years, but does it suffice today as many more questions remain unanswered? I flatly say "No."

CHAPTER 2
READER'S REACTION TO TRILLION THEORY

The preceding chapter offered a short preview of Trillion Theory. The longer version of Trillion Theory is available in 'Trillion Years Universe Theory' which was published in 2014. The remainder of this book focuses more on the vital role played by black holes as the builders of our cosmos.

This chapter deals with questions which have been posed to the author during his presentations at book fairs and science expos. Here are the most pertinent audience queries:

Question: Why should I give a tinker's damn? Does it really matter to me how our universe started or how it became so humongous? What's in it for me?

Author: Point well made. Though budgets were trimmed after man's achievement of landing on the moon, people still have an eagerness to explore space, the final frontier. Man's inquisitive nature is unstoppable. Ever since the caveman, humans have been on a long voyage of discovery. Each new century saw a faster paced world. Millennia passed between the early discoveries: fire, spear, and wheel. Thereafter, each great invention set the stage for industrial rallies: steam engines 1736; plastics 1862; telephone 1876; cars 1886; airplane 1903; television 1928; and microprocessor 1971.

What will man's next great invention be which benefits all of humanity? As a civilization, we face dramatic home-planet problems: overpopulation; lack of drinking water; and global warming. Will these escalating problems defeat man or will man be able to find world-saving solutions.

I'm betting man's next super-inventions will come from a greater understanding of our cosmos; therefore a greater knowledge of our own home planet. Will man be able to someday create new matter and elements and thereby build super-batteries to rid ourselves of the use of fossil fuels, or control the heat impact coming from our sun in order to defeat global warming? Or use these new elements to build flying saucers affording speedy travel between star systems?

Question: Why is your Trillion Theory important?

Author: My Trillion Theory states: *'The knowledge of how matter was made into the atoms and elements comprising the planets, moons and suns of our cosmos is an absolute integral part leading to the most prolific discoveries which a species such as man can ever hope to achieve. This hidden secret holds a vital key to understanding our universe.'*

So, human history shows that when a society lives under the incorrect theory, their progress is drastically held back. Illustrations can be seen with the old incorrect Flat-Earth Theory, wherein sailors venturing towards the horizon feared sailing over the edge into an abyss. Too, the incorrect theory of Greek astronomer Ptolemy, stating Earth as the center of the cosmos, held humans back for 1400 years.

As well, this author claims that belief in a Big Bang is holding us back from real eye-opening discoveries about our cosmos. For, if humans expect to ever truly conquer space, we need to uncover more of our universe's secrets than that supplied by an incorrect Big Bang theory.

Have you seen what people are saying? Several top flight scientists are finding fault with Big Bang. They look for new theory. *Small changes give rise to large consequences. And incremental changes can turn our understanding on its head.* So, we need to take a new gander at our universe, performing forensics, determined to find the truth.

Question: What can Trillion Theory explain in short form?

Author: The Trillion Theory preview in the previous chapter accomplished brevity. However, our cosmos is ultra complex.

While I would prefer to deliver my entire new theory to my readers in a lump sum, it takes substantially more pages than that. So my new theory reaches you chapter by chapter. I try to connect all the dots, for my new theory requires replacing some old beliefs with new universe ideas.

Since you might not accept my new theory at first glance, it'd be helpful to keep an open mind to my new concepts.

Question: What are some of these 'out-there' ideas?

Author: Here is a short list of my new Trillion Theory ideas:

• *Trillion Years* is the age of our cosmos. Much older than astronomer's estimates of 13.7 billion years which is just the age of the current rendition of stars in our sky.

• *Recycling* applies to our entire physical universe. My theory maintains that the solar systems of our universe (and their moons, planets, suns/stars) have recycled many times over with each cycle lasting approximately 15 billion years.

• *Black Hole prominence.* Trillion Theory claims that there is a black hole at the core of every sphere in our cosmos. These black holes are the building machines of our cosmos.

• *Light the main material.* Trillion Theory claims 'light' as the main material used by black holes to build the spheres, solar systems, and galaxies of our physical cosmos. We have yet to uncover all the marvelous secrets of light, and how Trillion Theory claims that light can be spun into matter. Up to this point in Earth history, no human has ever 'played creator' by taking light past its bending stage and forcing it to spin and coil into atoms of matter. Such a scientific discovery as the spinning of light into matter would be astronomical. Also, this discovery would provide proof for Trillion Theory.

• *Grew.* Our universe grew from small to its present gigantic size via recycling and reproduction processes which were built into the spheres of our universe.

• *Reproduction.* Grew and reproduction indicate some type of 'alive.' Yes, there is an 'alive' component to all the spheres (moons, planets, stars) in our cosmos. That doesn't mean there is actual life on the surface of every sphere. Till now, we thought of all the spheres as being dead-ducks. Now, to propose a 'life feature' inside the core of every sphere in our universe is an incredible claim. Rather, 'alive' means that all spheres in our universe have an 'alive black hole' at the core. This black hole provides many things to the sphere it occupies: the sphere's spin; the initial collection of matter inside and around the black hole to form the sphere; the gravity up to the surface and out into surrounding orbits; the reproductive method when the sphere dies; and the battling power to gain best position in the next solar system.

Question: I'm having a tumultuous time accepting a black hole at the center of Planet Earth, and our sun, and inside all the planets/moons of our solar system. Where's your proof?

Author: No proof thus far. However, if there isn't a black hole at the core of ever sphere, then Trillion Theory is way off base. But, a theory which places a black hole at the center of every sphere, and a supermassive black hole at the core of every galaxy, makes a whole lot of things in our universe suddenly make sense. It makes sense that a power play between battling black holes is the actual reason for the formations which result in solar systems and galaxies. My Trillion Theory first placed black holes at the core of spheres in 1998. Now in 2015, I am even more convinced that black holes are the builders, the operators, and the organizational recyclers of the trillions of spheres of our universe.

Question: Are there noted astronomers or astrophysicists who support your view of a black hole inside every sphere?

Author: Although my Trillion Theory began back in 1998, it is only in 2014 that it was published and only in 2015 that I word began to spread. Although most astronomers likely have no knowledge of my theory as yet, many astronomers show signs of a movement in a Trillion Theory direction.

For instance, on the internet go to a documentary video called *Secrets of a Black Hole by HD Universe Channel.* This video was released January 18 of 2015. The things which it says about black holes fit perfectly into my Trillion Theory. Earlier astronomers had thought that a black hole was just theoretical or extremely rare. In 2015, a HD Universe Channel video said this about black holes: *'Black holes are everywhere; all over the universe; millions of them. A supermassive black hole seems to be at the center of every galaxy in the universe. And it appears that supermassive black holes can do more than just control an entire galaxy. A supermassive hole also has the ability to build a galaxy.'*

Now, scientists are recalibrating their respect for black holes – seeing black holes are constructive builders, not just destructive monsters. Penn State astrophysicist Yuexing Li had this to say, "Now we see black holes were essential in creating the universe's modern structure."

In 2003, a sky survey showed giant black holes centering most, perhaps all, galaxies. Most of these giant black holes were found to be 13 billion years old. Since then, researchers have tried to figure out the origin of these primordials.

Trillion Theory takes that a step further by declaring that there exist two types of black holes in our universe. Yes, the supermassive black holes at the center of galaxies are the supermasters, while all less large (lesser black holes) exist at the core of every star, planet and moon in our universe.

Question: Okay, so who built all these black holes?

Author: That's difficult to say at this stage. The same could be asked of the supposed Big Bang: who or what supplied the matter for Big Bang's supposed explosion? However, once we can solve more of the mysteries of our physical cosmos, we'll be better positioned to tackle such a query.

Question: As a well informed follower of physics, it has been the understanding that the amount of matter in our cosmos is limited, set at the moment of origin. Matter can neither be created nor destroyed. So, black holes wouldn't be able to bring new matter into our universe to build more spheres.

Author: An incorrect belief. Trillion Theory shows that our universe has a limitless supply of energy material provided to it from the outer ocean of static light, and this energy is continually pulled into action by black holes and utilized to increase the amount of matter available for black holes to build and increase the sphere population of our cosmos.

Question: What is the purpose or the goal of your new cosmology Trillion Theory?

Author: The primary goal of Trillion Theory is to provide a NEW PARADIGM and POWERFUL NEW VOICE to present a thorough depiction of how our universe really began over a trillion years ago; and thereby how our universe originated, grew, and now operates. A second aim is to identify possible evidence as PROOF. Lastly, to replace old Big Bang which incorrectly depicts our universe to be only 13.7 billion years; just the age of older present stars in our sky.

Question: Where does pure science fit in?

Author: Later in this book I introduce the incredible engine (which creates gravity) inside of a black hole. A black hole within the core of a sphere has such intricate working parts that it must have been purposefully designed by a mind(s) with absolute incredible super knowledge of science.

Question: You say universe's creator was a scientist?

Author: I prefer to say, 'used incredible science,' rather than 'created out of nothing.' I don't tackle the idea of a creator or artisan. It will be quite a time after we fully understand our physical universe that we will be better equipped to take on such questions as 'who' or 'what' used scientific genius.

I firmly believe that tremendous pure meticulous science was behind the construction of our cosmos. It was Stephen Hawking with his books *A Brief History of Time* and *The Grand Design* who brought more of a scientific approach. But, a renowned Hawking was also fooled by Big Bang, often trying to push a square Big Bang through a round hole.

Trillion Theory takes a hard fast step at explaining the techniques involved in the science behind the building of our physical universe. Further, Trillion Theory places black holes at the forefront, showing how their discovery was just the initial step in uncovering the fantastic job black holes have done in building spheres, solar stems, and galaxies.

In Trillion Theory there is the use of the words 'clever' and 'strategies' when describing the ultra-scientific design, construction, and operation of our physical universe.

Question: Astronomers state that light could never ever escape from a black hole. You say otherwise.

Author: I hate to say it, but astronomers are still stuck with Big Bang thinking. However, I see signs of them questioning those beliefs. When astronomers first discovered black holes a few decades ago, they were incorrect to presume that a black hole was a death trap for light. They wrongly stated, 'light can never escape from the clutches of a black hole.'

For, while a black hole does pull in and entrap light for anywhere up to billions of years, that is not the end cycle for the light. Our 10 billion year old sun, for example, has a black hole at its center which billions of years ago consumed light and entrapped light as matter around itself. But now, aging in its cycle, the sun is in the process of losing control of the tons of light which it consumed, and that matter is unraveling back to light escaping from the black hole sun.

Astronomers see Supernovae explosions of stars; they see a star collapsing hard, creating a black hole. They presume that the exploding-then-imploding star creates the black hole. Whereas, Trillion Theory says that the black hole was originally the builder of the star, the entity which built that star around itself in the first place. The explosion of that star's contents, after billions of years of entrapment, exposes the naked black hole which was originally the star's builder. The building of the star around the black hole was the early stage of its cycle, followed by the 10 billion years or so the star spent slowly unraveling matter back to light.

Question: How can our universe be a trillion years old when our solar system is a young 8-10 billion years of age?

Author: Our solar system and the youthful stars which are in our sky are just the latest current rendition. They are far younger than the trillion year history of our universe and they are just the latest episode in a long history witnessing incredible recycling of our universe from small to gigantic.

An analogy to our universe would be a forest where we only see current trees. However, the forest's history has seen thousands of tree generations deeply buried. Similarly, our universe's history has been lost by the recycle of its spheres.

When the spheres of a solar system eventually die after a lifespan, their graveyard quickly recycles, growing into the next new solar system. Trillion Theory maintains that the universe solar systems, each comprised of moons, planets, suns/stars, recycled many times over with each cycle lasting roughly 15 billion years or so, with 67 such blending recycles culminating in an overall trillion year universe age.

Question: Some people say that black holes are portals to other universes or openings to tunnel systems which would allow us to travel our universe?

Author: That would be terrific, if true. And, I do believe that one day it will be possible to far eclipse the speed of light and travel a super highway across distant space, or just leave here and zap there. But I don't feel black holes are those portals; however, they could help us in those discoveries.

Question: So, do you think our universe was purposely built or just a big accidental occurrence?

Author: My hat goes into the ring with purpose. Each day we find our cosmos to be more ultra complex. Our cosmos wasn't just a happenchance. But, on the other hand, we have been too simplistic in our beliefs. 'Oh, God created that.' No, God didn't just create that, rather some incredible scientific mind(s) designed and built this universe with clever intent. As we discover more about our cosmos, we will find more cosmic rules which pertain not just to our solar system, but everywhere across the wide expanse of our cosmos.

Astronomers have spotted in far corners of our cosmos that hot suns are always the central largest sphere of a solar system. Happenchance can't explain this same configuration throughout our cosmos. Thus, something else is afoot, much beyond the happenchance of a Big Bang or Nebular Theory.

Therein, on a clear night, when I gaze up at the evening stars, the scene amazes me each and every time. Planets and distant stars look so pristine, so clear cut with vast amounts of space in between. For, the formula deployed to build all this was amazing. It excites me to discover more. And always we're provided with clues: a rainbow; a bolt of lightning; fire; and a supernova – all stirring man's voyage of discovery.

Next chapter: 'In furthering our understanding of black holes, we gain a far greater understanding of our cosmos.'

CHAPTER 3
BLACK HOLES ABOUND

Weird mysterious black holes.

They are the least understood objects in all the cosmos. Prior to my Trillion Theory, no one had been able to explain their exact role in our cosmos. Prior to now, black holes were always thought of as exotic bizarre monsters in our cosmos. Only lately are there astronomers also seeing black holes as entities which bring order and organization to our cosmos. These astronomers are converging with Trillion Theory, even though they have never yet read my new theories.

The main reason that black holes are misunderstood and viewed as being so mysterious is that they are beguiling presenting different looks depending upon their size and the phase which they are in. Each phase shows a different appearance. Astronomers often see dark naked empty black holes ready to attract, devour and then spin light into their bowels; or often, as bright neutron stars surviving after the Supernova collapse of a star; and often, as the gigantic supermassive black holes at the hub of spiral galaxies.

What astronomers never get to see are all the other undetectable black holes cloaked inside of and at the core of every moon, planet, and star. Trillion Theory says that a cloaked black holes is inside of each and every sphere in our universe, regardless if that sphere is a moon, planet, or sun.

Yet, proof of Trillion Theory is still distant.

Of all various the hypotheses proposed by others, it is only Trillion Theory which states that light traveling close to a naked black hole is bent, curved, and then finally **spun into dense matter** inside of the naked black hole. As the hole devours more and more light to spin into matter, a sphere is built around itself by the hole. The mechanics of this process are: naked hungry black holes possess an ultra-fast spin resulting in a powerful gravitational field around themselves. This spin warps space immediately surrounding them such that anything traveling into that space, such as a beam of light, is slowed, bent, curved and then swallowed up by the black hole. Trillion Theory says that swallowed light is spun into atoms of matter which are then pent up and stored inside of the body of the sphere which the black hole builds around itself. As the black hole eats to quickly and fills, it changes its duties from eating to control of its contents. Example: Planet Earth, with a black hole at its core, is presently in a long control phase, attracting new light with gravity but having no more room to spin light into matter.

Thus, Trillion Theory discovery dramatically changes how we should think about the origin of the moons, planets, stars, and galaxies of our universe. Trillion Theory states: *'**Black holes take light and spin it into matter, holding it tight for up to billions of years. Therein, black holes are the spinners of light into matter inside of and around themselves to build the very spheres which exist in all the galaxies comprising our cosmos. Black holes were (are) the builders of today's cosmos.'***

Thus, Trillion Theory discovery helps to unravel some of the complex mysteries of our universe by focusing on black holes, with their major role in the origin and growth of our universe. They have brought phenomenal workmanship.

History of black hole discovery.

To put it quite bluntly, tons of scientific work-time has been spent studying and theorizing about black holes, yet they are still barely understood to this day.

The concept of a body with such great gravity that light could never escape was proposed way back in 1783. Then in the early-to-mid 1900's Albert Einstein, and others after him, continued to pursue an interest in the idea. 1964, Ann Ewing used the term in her article *Black Holes in Space.* However, the credit for naming black holes goes to John Wheeler in 1967 as he coined the phrase. Thereafter, that name stuck and black holes became mainstream research subjects.

Even then, black holes were still regarded as theoretical curiosities. However, with the discoveries of neutron stars, pulsars, and quasars - black holes gained traction.

Neutron Stars: Small but mighty, possessing rapid spin. They are the remaining birthing cores of massive stars which went Supernova and collapsed. The surviving dark black hole retains the gravity spin to attract huge amounts of light in a rush back into the hole, thereby showing a bright glow. A pulsar is a Neutron Star seen blinking on and off due to the pulsating view caused by the stars rotation.

Quasars: A Quasar is the supermasssive black hole centering a galaxy, plus two twin jets pluming from the poles, plus the array of bright surrounding stars outshines a galaxy.

Naked phase of black holes is extremely hard to see.
In 1972 an astronomer pointed his telescope towards a dark spot in open space detecting radio emissions. He didn't know he was looking at. Finally, after months, the dark spot in space eclipsed an adjacent star partially blocking out its light. An actual naked black hole had been discovered.

Gradually astronomers found more about black holes.
Some black holes were found to be the survivors after the collapse of massive stars which ran out of fuel near the end of their life cycle, then went Supernova, and then finally collapsed down into themselves. Plenty was done to classify the sizes of various black holes. These black holes sizes were dependent upon the star mass from which they had come.

The notion that black holes sucked in everything was correct. The area directly around the black hole was named a black hole's Event Horizon – which was the boundary that the black hole fashioned around itself in space. Anything passing into this event horizon was pulled inward into the black hole. It seemed that the black hole actually had the power to deform the space directly surrounding itself. Not even fast paced light racing through space could escape the clutches of a black hole. The swallowed light appeared to disappear into lockup forever. Plus, no information about that light was ever again received by the world outside of the black hole. In that black hole's Event Horizon, there were no paths which led away.

Shape-wise, this gravity boundary surrounding the black hole was found to be oblate – meaning that the boundary extending out from the hole's equator was greater than the boundary extending out from the flattened poles. We see oblate with Saturn's equatorial rings; and the linear shape of our solar system; and the pancake shape of spiral galaxies.

Further out from the black hole, an extended area was called the Ergosphere. This area extends well out past the black hole's event horizon and is shaped oblate (more linear than spherical). In this area, it is impossible for things to stand still. Example: In Trillion Theory, the powerful black hole at the center of our sun extends its Ergosphere out to all the planets/moons which cannot stand still, but rather must orbit the sun in the direction in which it is spinning. Thus, this power that the sun has to drag its Ergosphere space around itself also drags all the spheres in our solar system into orbit. Note: It is possible to escape this Ergosphere if humans were ever intend to do so.

Comparing the 'Event Horizon' and the 'Ergosphere' of a black hole as Trillion Theory sees them:

Both are important ramifications surrounding black holes. The 'Event Horizon' is smaller and closer in around the black hole. Any light passing into the Event Horizon would be attracted, bent, and finally spun inwards towards the black hole where that light would be spun into matter to be locked away for up to billions of years. Once full on its inside core, the black hole would then add to its periphery by spinning even more light into matter to form a sphere of matter around itself. After the black hole is totally full inside and on its surface, the Event Horizon can still attract light and bend it, but it can no longer spin that light into matter. Example: Planet Earth's black hole is full.

The black hole's 'Ergosphere' extends much further outwards from the black hole than the Event Horizon. The Ergosphere is the area around a black hole where objects must be in motion. Our sun's Ergosphere extends outwards to the further planet of our solar system. All the planets and moons within the inner gravitational swing and pull of the sun's Ergosphere must be in continual motion revolving in orbits around our sun. This Ergosphere can control the smaller black holes in the vicinity of a larger black hole when all are naked and feeding on light. And this Ergosphere remains in place after all the black holes of an area become full as moons, planet, or a sun. The sun, which has the largest Ergosphere, controls moons and planets in orbits.

The speed within both the event horizon and the Ergosphere is greatest when the black hole is naked and feeding to spin light into matter. These speeds slow to slower more sluggish once the black hole

Question posed by the reader: If nothing can escape the event horizon which is close to the black hole of our sun, how come light can escape from our sun.

Answer from the author: The black hole at the center of our sun completed its eating of light billions of years ago.

Since getting full, the sun's black hole has just been trying to hold onto its spun contents for the past billions of years. Now, it is losing that battle as its contents loosen and eventually escape as light from its surface. So in fact our sun's event horizon no longer exists while its control over space in its Ergosphere still is in effect as its strong gravity plus great mass project a gravitational pull outwards to the planets and moons surrounding it. Also, light which escapes the sun can also find its way out past the Ergosphere to travel the universe. Note: Trillion Theory disagrees with those who say that light never escapes a black hole. Trillion Theory shows that a black hole entraps light for billions of years, but light is always destined to escape no matter how long it takes since no sphere (moon, planet, or star) can last forever. Spheres always age-out (lifespan generally 15 billion years) and then their matter always escapes back to energy.

Nothing can escape a naked black hole, True or False?

To astronomers, a black hole devouring light seemed like a forever thing. Nothing could ever escape a black hole once it was swallowed up. It was as if a vacuum cleaner had sucked an object up forever. But in 1974, Stephen Hawking made his most famous discovery: black holes emitted radiation, meaning there was indeed something capable of escaping a black hole's clutches. And, furthermore Trillion Theory predicts that light always eventually escapes the confines of a black hole, most often taking billion of years.

But, nothing about black holes has ever been easy or quick for astronomers (astrophysicists). Each time they seem to have black holes figured out, a new fire-eating dragon rears its ugly head posing problems to solve.

Even physicist Stephen Hawking recently declared that black holes exist differently than first thought. Thus Hawking re advocated that while light can't escape from a black hole, they are sort of stuck or stored in a holding pattern.

This author likes the idea that Hawking is beginning to think more like Trillion Theory where light is spun into a black hole, and is stuck inside of the hole for billions of years inside of a planet, moon, or star, ready to one day eventually escape when a Supernova occurs. The Supernova blast frees all the pent up light stored as atoms inside of the spheres of that solar system.

One certain, astronomers and astrophysicists are looking for the next mind-bending property of black holes. Trillion Theory supplies many 'next ideas' which contradict some of the past accepted black hole views.

Let's ask some pertinent questions about our universe. To do this, let's step back and start from scratch, with open minds, and no theory such as Big Bang to direct our thinking. Here goes:

- Why is our universe so gigantic? Why not small?
- Why island solar systems and island galaxies?
- What controls formation of solar systems and galaxies?
- Why is linear formation shape seen all across the cosmos?
- Are there rules or laws seen across the cosmos?
- Why is space so vast between these galaxy islands?
- Why is space on its own so absolutely nothing?
- Why are planets cold, while stars are so hot?
- Why do planets, moons, sun spin on their axis?
- Why do galaxies spin around a supermassive black hole?
- Why is spin seen in everything, atoms to galaxies?
- Why does this spin seem to go on forever?
- What created the pull of gravity?
- Why are there black holes in our universe?
- Why do black holes spin at thousands of rpm, or at all?
- Why are black holes now known to be in the millions?
- Why supermassive black holes at galaxy centers?
- Why star crowds around a supermassive black hole?
- Why a black hole survivor after a Supernova?
- Can light ever escape from a black hole?
- In a dark universe, why is there light?
- Which came first in our universe, light or a star?
- Why does an atomic explosion send out light?
- Does time exist across our entire universe?

These are a few of the most pertinent questions. Answer these questions, and you are close to unraveling the yet to be discovered secrets of our fascinating universe. Trillion Theory puts forth answers to these questions, offering a new opportunity to think outside-the-box of Big Bang, and to view our entire universe in a new ultra-modern fashion. My advice is: think from a different perspective; from a strategic view. For Trillion Theory demonstrates how clever strategy was a big part of how our universe was designed. Black holes are a prime example of this strategic style.

What new views are presented by Trillion Theory?

- Trillion Theory professes that our universe is far older than 13.7 billion years old. Rather a trillion years old.
- Trillion Theory claims our universe grew to its enormous size of 73 quintillion stars - no Big Bang at the origin.
- Trillion Theory denounces happenchance. Definitely, top scientific skills were deployed in the specific design and construction of the 'black hole building machines' as the 'recycling entities' of our universe.
- Trillion Theory places a black hole at the center of every moon, planet, star (sun), and galaxy in our universe. These various black holes vary in size.
- Trillion Theory states spheres and solar systems comprising the galaxies of our universe recycle roughly every 15 billion years. Each cycle ups the count of stars and solar systems.
- Trillion Theory predicts that there exists an endless supply of energy available to grow our universe to infinity. Thus, the number of solar systems, stars, and galaxies in our cosmos will increase exponentially forever.
- Trillion Theory (TT) states that a small galaxy can totally recycle like a solar system, but generally large galaxies see the supermassive black hole at its central bulge take control of that galaxy for hundreds of billions of years.

- TT emphasizes the incredible properties of light as the absolute energy units deployed in the recycling process.
- TT proposes that black holes possess reproductive (terms duplication and replication are used). Their means of replication is different than any seen on Earth, yet black holes are akin to a living growing entity, even though they are more like machines. Black holes come programmed with instructions enabling duplication after a Supernova destroys a sun and its solar system. This growth feature perpetuates and grows the number of black holes available to spin light into matter for a new solar system.

This last feature given to black holes by Trillion Theory may seem tantamount to science-fiction. Yet, Trillion Theory likes to show at each and every step that our universe is all about the very tops in science. Black holes 'fit the bill.'

What do some modern astronomers have to say?

A group of top astronomers do work at the Mt. Laguna Observatory in California. Their main jobs are to study black holes of all types and sizes. Their modernistic thoughts about black holes are most interesting. These astronomers, were educated as to the following about black holes:

- black holes were thought to be a rare cosmic feature.
- black holes were difficult finds in the dark of space.
- black holes fostered terrifying violent drama.
- black holes brought only chaos and destruction.
- black holes were the most destructive cosmic objects.
- black holes could rip apart a star.
- black holes could make a star go Supernova.
- black holes were a one way street for captured light.
- black holes could swallow all nearby light and matter.
- black holes could spin furiously.
- black holes all spun in only one direction.

Now, these top astronomers at Mt. Laguna Observatory are unlocking more of the secrets of black holes:
- black holes are not a rare cosmic feature.
- black holes are becoming easier to find in dark space.
- there are billions of black holes in our cosmos.
- the fact there are so many black holes is extraordinary.
- black holes are the most mysterious cosmic objects.
- so far, black holes defy understanding.
- black holes are not just chaos and destruction.
- black holes are not an eternal one way street for light.
- black holes can spin in either direction.
- black holes can spin at millions of miles per hour.
- black holes stage the most violent battles in the cosmos.
- black holes also have another gentler side.
- supermassive black holes are at the center of galaxies.
- these supermassive act as anchors for a galaxy.
- supermassives control the organization of a galaxy.
- black holes have a purpose of being cosmic organizers.
- black holes seem to shape our universe.

Trillion Theory takes new findings a step further:

Trillion Theory says we need to do a makeover and get over the preconceptions brought by Big Bang. It's time to see black holes as builders, organizers, and controllers of the contents of our cosmos. Black holes are machine (precision-like) builders of galaxies, solar systems, and spheres.

Stars don't just randomly occur across our cosmos. They exist within humongous organized islands in space, unit islands called galaxies. And within these galaxies, there exist smaller organized units known as solar systems.

Trillion Theory proclaims that the unit galaxies and unit solar systems and all the spheres, were built and organized by the machine-like workers known to us as black holes. These black holes built this much-ordered cosmos system.

CHAPTER 4
BLACK HOLES
OUTSIDE AND INSIDE

No one has yet ever seen the real inside of a black hole. Obviously, this is the most mysterious and possibly the most dangerous spot in our universe to a human. We can only fantasize about entering into a black hole, and that fantasy might have horrific consequences. The interior zone of a naked black hole possesses the most powerful force and fastest spin in our universe. It takes something as universally indestructible as 'light' to survive such a passage.

This chapter first describes the outside of and then the area around a black hole which astronomers see when they study a black hole. Then, we will go inside a black hole to see what has never been seen before; with the purpose of determine the *engine structures* which cause the events in the vicinity of the black hole.

Getting a view of a black hole – thus far there seems to be no way to see cloaked black holes.

Trillion Theory advocates the presence of a cloaked black hole, under-disguise, at the core of every sphere including Earth. The one residing at the core of our planet is hidden away under far too many tons of matter to be seen. The only time this black hole will uncloak is when our planet dies and sheds all of its matter, thereby exposing the black hole.

Except for the obvious permanently-visible uncloaked gigantic supermassive black holes at the center of galaxies, most of the lifetime of all other black holes is spent cloaked and concealed inside of a sphere, masquerading in various sizes as a moon, planet or sun.

Outside view of a black hole.

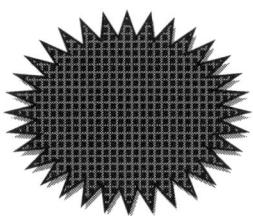

The naked black hole appears inky black, spinning at great speed on its axis. Its rough surface provides it with better grip for pulling nearby light inwards.

The area surrounding a black hole.

When astronomers study black holes, here is what they see as shown in the graphic. The inner black area is of course the *black hole* itself; absolutely dark and spinning at tremendous velocity. Trillion Theory refers to this black hole as a naked black hole when it is feeding and pulling light into itself. Shape-wise the black hole is very slightly oblong, namely slightly wider across at the point of its equator. This results from its great spin speed and the centrifugal force which happens around its axis which runs pole to pole.

Directly surrounding the naked black hole is the *Event Horizon* where a black hole displays its strongest influence. *Event Horizon* space is bent inwards. Any light passing into that locale is curved and pulled into the black hole.

Trillion Theory says that the light is taken past curving and bending, to the point of spinning into atoms of matter. This differentiation (made only by Trillion Theory) has huge repercussions as physicists studying light have never yet been able to spin light into matter. Trillion Theory states that it is possible to spin light into matter and shows how black holes actually are the engines capable of the feat.

Outside the *Event Horizon*, extending quite far out into space is the *Ergosphere*. The outer limits of this Ergosphere depend directly upon the size and power of the black hole. In Ergosphere no object can stay still, as all objects must be in motion. This continual motion is caused by the powerful spin of the black hole. It drags the Ergosphere around itself. Plus, all the spheres within the Ergosphere are dragged around in orbits within Ergosphere's gravity zone.

Let's first examine a *spiral galaxy* as an example of a black hole model, and see some slight variations.

The *supermassive black hole* at the hub of a galaxy has evolved, displaying some different characteristics compared to a small naked black hole. But, many influences remain the same. A supermassive black hole can still pull light into its *Event Horizon*. which is filled with massive stars held close. The *Ergosphere* (gravity) of the supermassive black hole can extend billions of miles out into the spiraling arms of the galaxy. (Supermassive black holes, builders of spiral galaxies, are discussed in greater detail in later chapters).

Next, we examine *our solar system*. Trillion Theory states 'that there is a cloaked black hole residing at the core of our sun.' This black hole was naked in early days. But as it spun light into tons of matter around itself, the black hole became hidden from view. Then, over billions of years, the black hole over-filled as a result of its gluttonous eating. It became sluggish with lower spin speed, and its matter began to unspin back to light, thereby turning it into a star. Since it is still too full, its black hole continually weakens in regard to maintaining control over its contents, this light in always escaping from the surface of our sun. However, for now, the black hole at the core of our sun adds the sun's mass to its own power to still exert a powerful and distant Ergosphere (gravitational force) to hold all the planets and moons of our solar system in continuous orbital motion.

Lastly, we examine *Planet Saturn* in our own solar system. Trillion Theory states 'there is a cloaked black hole residing at the core of Saturn and a black hole at the core of each and every planet/moon in our solar system.'

For Saturn, this black hole was naked in early days, but as it spun light into tons of matter around itself, the black hole became hidden and unseeable inside. Then, over billions of years, the black hole at Saturn's core became overly full making the black hole sluggish with lower spin speed. Yet, Saturn's black hole is still able to hold matter on its surface. Saturn can attract light into its event horizon, it can even bend and refract this light, but it is too full to spin any more light into matter. Within this Event Horizon are the Rings of Saturn, but the thick matter surface around Saturn protects the rings from falling into the black hole. The Ergosphere (gravitational field) of Saturn extends quite away outwards as the planet holds 62 moons in orbit, with Titan by far the largest. Note: Evolution-wise, Saturn has become a gas giant, meaning that its black hole has lost some of its hold on the matter surrounding it. So Saturn is less solid than Earth, but as gaseous as a star. It may someday evolve into a star, but more likely it will have its matter destroyed when our sun goes Supernova meaning that the black hole at the core of Saturn would survive and return to being a naked black hole eater and spinner of new light into new matter.

Inside, into the bowels of a black hole.

Now we move inside a black hole – to an area never seen before by human eyes. What could be inside?

The interior zone of a naked black hole possesses the most powerful force and fastest spin in our universe. We go inside of the naked black hole to see what is causing events to happen surrounding the black hole. This is vital to Trillion Theory which bases its growth model for our universe by the events which are predicated within and around black holes.

Astronomers who study black holes, readily admit today that supermassive black holes use their powerful gravity to be the organizers and controllers of galaxies. Trillion Theory further states that black holes are more than just control organizers, they also were the machine-like entities which built spheres, solar systems and galaxies in our cosmos.

Inside, what might be the structure of a black hole?

What is it about the structure of black holes which makes them able to be the controlling organizers of solar systems or galaxies in our universe?

Trillion Theory advocates that regular black holes (those inside of moons, planets, and suns) are slightly different inside than the supermassive black holes which are at the hub central bulge of galaxies. Those slight differences will be shown more in later chapters.

For now, we are only able to strategically envision the structure enabling a black hole to spin light into matter and

then keep that matter spinning as atoms for eons. Thus, we can only guess at the absolute exact inside structure of a black hole. However, knowing what a black hole does, and the powers it possesses, it is possible to deduce the type of internal structure which a black hole would need to have. Here is the possible internal make-up of a naked black hole.

Black Hole Structure

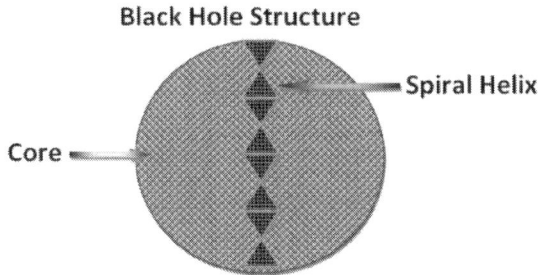

Core ◄—

Spiral Helix

A Black Hole in its naked phase. It has a spinning 'Spiral Helix'
running through its middle and a 'Core of Compartments'
(each spinning) as main parts of its roundish design.

The reason a black hole achieves such a high spin rate is
the structure of its axis. Compare to a spinning toy top.

The pump toy top uses a pump handle to send a fast spin to its
body. The harder and faster the handle is pumped, the faster
the body spins. As the handle depresses down, the body is
imparted a spin, as the handle raises the body rests in spin.

A main key to black holes is their *pumping helix axis*. It
works similar to the twisted rod at the center of a small toy
called the *pumping toy top.* The handle of the toy pumps
down and then up, moving a spiral rod in the center of the
toy, creating spin, and the spiral rod axis then imparts this
spinning power to the body making the toy top spin on the
floor. Linear spindle motion torques into circular spin.

In a similar manner, the helix axis of a black hole uses its
elastic properties to lengthen and shorten the helix axis to

cause the core to rotate. This can go on to perpetuity as a shortened helix has a need to lengthen and a lengthened helix must shorten. Interestingly, it is the same type of pump action which will occur inside of an atom as described later in this book. That's why an atom can spin for eons even on Earth where there is resistance - that internal pump action.

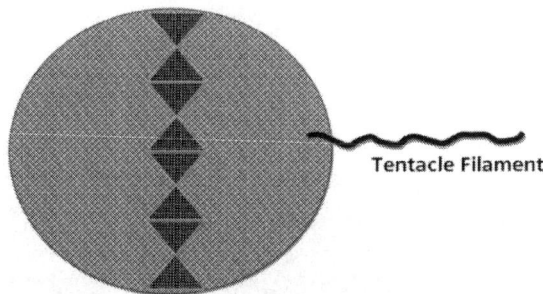

Tentacle Filament

A *tentacle filament*, as one example of millions, extends from a compartment of the interior body of the black hole. These myriads of filaments utilize their grasp to aide in the securing and pulling of light inwards into the core of the black hole.

An important part of the structure of a black hole is its tentacle filaments. Every compartment has one. When a black hole spins at a rapid rate, the tentacles extend outward from each one of the compartments to grasp onto passing light, pulling the light inward to be spun into matter.

Inside of a naked black hole, processes are occurring.

The black hole begins its new life as a naked empty black hole spinning at a fantastic unabated speed. The core body of the naked black hole, with its myriad of compartments, spins around its rod-like spiral helix. Phenomenal elasticity of the spiral helix pumps up and down, imparting a spin to the spiral rod. This spiral helix then imparts this motion to the body making the entire black hole pivot on its axis. Linear spindle motion torques into circular spin of the entire body. The naked black hole is now ready to begin feeding.

The fast spinning motion of the black hole projects a strong pull into the *Event Horizon* surrounding the black hole. Any light contacted within that Event Horizon can be pulled inwards, bent, and finally spun into matter inside of the black hole.

Light is pulled inwards from the surrounding *event horizon* into the *core* of the naked black hole. The first attracted light will make it all the way to the spiral helix and the very center of the black hole. In Trillion Theory, the laws pertaining to black holes show that the spiral helix will fill up first, able to grow its length. Note: Remember, this is evolution and survival of the fittest. The black hole is programmed to become as large and powerful as possible in competing against all other black holes in the universe.

Next, the black hole will spin light into matter to fill the compartments of its main interior body. With this fill up, each compartment utilizes its elastic feature to stretch and become larger. At a maximum stretch, the compartments can subdivide something akin to a living tissue thereby doubling several times over its number of compartments. With filling, stretching, subdividing techniques, the black hole adds to its entire structure giving its spiral helix greater length and girth; plus adding even more body to match the new length and girth of the helix.

What happens when once-naked black hole has filled up its entire interior with matter spun from light?

Trillion Theory shows that each and every black hole will continue to spin more light into matter, even after it has totally filled its interior. The Event Horizon around the black hole continues to attract more light. However, the naked black hole is no longer naked. It is becoming fully dressed and cloaked as it spins tons of light into a body of matter to surround its interior. As this process continues, the black hole builds itself into a sizeable sphere such as a moon or planet. The size of any such resultant sphere will depend upon the size of the black hole which is doing the building. Other factors which determine size would be the availability of light to spin into matter, and competition from other naked black holes for that available light.

As the black hole becomes totally full with matter, it can still attract light to its Event Horizon, but the black hole can no longer spin that light into matter. Examples would be any of the moons or planets of our solar system which can attract and even still bend light towards their surface. However, the speed of spin of the full black holes has slowed and they no longer possess sufficient power to spin light into more matter.

Upon becoming full and heavy in weight, the black hole would use its spin and the weight of the body of matter which it has built around itself to extend it *Ergosphere* (gravitational pull) far outwards from itself. Any other spheres, smaller compared to the newly built sphere, which happen to be quite nearby, might find themselves trapped within the *Ergosphere* (gravity pull) of the newly built sphere for billions of years as they go into orbit around that sphere.

What happens if more than one naked black hole is at the same time filling up in the same locale?

Now, let's imagine that there are many black holes located in a certain zone of our universe. An example would be the zone where a new solar system is just about to be built. There are anywhere from 5 to 150 naked black holes. They are the naked survivors of the Supernova which destroyed their solar system and thereby melted and released back to light all of the matter which had existed on the previous sphere bodies.

All of these now naked black holes will engage in a battle for the available light in that zone. Survival of the fittest (size, speed, and best location) will determine the final results of that battle. Those results will follow the *laws of naked black holes* and *the laws of spheres* in determining the final organized make-up of the new solar system in that particular zone of a galaxy.

Summary of Black Hole Structure and Purpose.

- Black holes are the engines of our cosmos
- Black hole is round (slightly oblong) in shape
- Black hole compartments surround a spiral helix
- Black hole structure designed for fast rotation
- Black hole rotation begets an event horizon
- Black hole event horizon attracts nearby light
- Black hole spins light into atoms of matter
- Black hole fills its interior with spun matter
- Black hole adds more atoms to its body
- Black hole forms sphere of matter around itself
- Black hole spin rate slows once it is full
- Black hole core spin rotates the moon/planet/sun
- Black hole gravity holds other spheres in orbit

Note: Provocative thought has always led the way. Science runs in to later prove or disprove. 'No great idea ever started without first being thought of as outrageous or ridiculous.'

CHAPTER 5
AGELESS TRILLION
YEAR UNIVERSE

Declaration by Trillion Theory: Our cosmos, at a trillion years of age, is much older than the present 13.7 billion year estimate. Far older than the stars we see in the present sky.

Using powerful telescopes, astronomers have catalogued some of the oldest stars in our universe to be in the 13 to 14 billion year range. From this viewing evidence, they estimate that the age of our universe is 13.7 billion years because that's the age of these oldest stars. They do not realize that the universe knows how to reseed itself from old dying stars to recycle over and over again. And Trillion Theory says that our universe has been doing this for a trillion years.

Age crisis in the cosmos.

In 2007, a unique star was observed to be 18 billion years old, providing a puzzling quandary for astronomers as to how a single star could be older than their supposed 13.7 billion year old universe. A similar paradox continued in 2013 with the discovery of the impossible Methuselah star, estimated at nearly 16 billion years old.

The Answer - Can't see the universe for the stars.
A Comparison to: Can't see the forest for the trees.

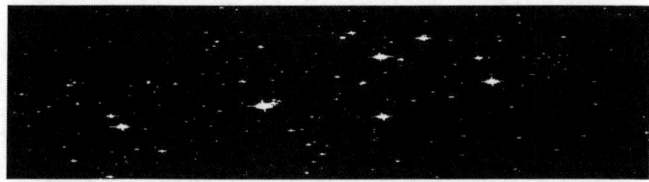

A great analogy can be drawn between the trees in a forest and the array of stars in the cosmos. Upon entry into a forest, an individual sees trees of various ages: old trees and young saplings, with the very oldest trees in this particular forest two to three hundred years old. Without experience, it is easy to incorrectly conclude that forest is a ripened three hundred years of age. Yet, that forest has recycled hundreds of times and is millions of years old. Proof is found in the old dead trees rotting away deep below the forest's floor.

However, the ancient history of our cosmos is well hidden within the recycling process of its spheres. Stars live so much longer and age so slow, that humans totally misread the stars. Like a tree, stars also have a recycle life. Stars experience a black hole birth followed by long life with a general upper limit of 15 billion years. At death, the surviving black holes grow in number as they begin their next 15 billion year cycle. Therefore, star populations of galaxies always increase with each 15 billion year recycle. Stars, planets and moons die and then perpetuate as new spheres in the next cycle of their area of the universe, making for more and larger galaxies and solar systems. Overall the number of stars grows with each new cycle.

People incorrectly miscalculate our universe's age using the oldest present-living stars as their gauge. Unfortunately, no star graveyard has left markers behind as evidence. Old dead stars so wondrously and completely recycle leaving no seeable traces (except for naked black holes ready to form the next solar system in that locale). Remnants of the ancient stars and solar systems are near impossible finds. Each old star cycle is lost to history, while each new cycle appears unique providing the illusion of being the one and only solitary cycle.

Trillion Theory says that we are presently living in the 67th of the 15 billion year cycles of our universe. But, for us there is the illusion that we have been the one and only cycle.

Star population of our universe.

In this 67 recycle of the stars, to approximate the number of stars, multiply two hundred billion galaxies (approximate) by three hundred and seventy million stars per galaxy = over 73,000,000,000,000,000,000 (quintillion) stars. Trillion Theory uses this tally, while present day astronomer estimates have placed the number even larger.

Size our Universe.

Our universe is trillions of miles in diameter. An easier way to explain that distance is in light years, which is the distance which light travels in a single year. It takes eight minutes for light to get from our sun to Earth, so light travels a long way in our universe in one light year. Yet, our universe is far too big for light to move very far across its expanse in a single light year. Our universe has grown each recycle to being hundreds of billions of light years across.

Correlation between population, size and age.

During universe history, stars increased in number each recycle, as did the spatial size of the universe. This took time, hundreds of billions of years. Star population, spatial size, and universe age are all intertwined and interdependent. But, our cosmos doesn't really show its true trillion year age, hiding behind the present cycle of stars.

Age of our Universe.

While Trillion Theory calculates our cosmos at a trillion years of age since inception, that estimate is based upon growth starting from one solitary star. However, Trillion Theory concedes the possibility that initial onset of our universe may have begun with hundreds or thousands of initial black holes forming into stars. Therein, such a faster start could have taken 100 or 200 billion years off the age.

However, Trillion Theory mathematical calculations begin with just the one star and grow from there with each recycle.

The following mathematical progression shows doubling of the star population of our universe each 15 billion years. The numbers start from 1 star in cycle 1, doubling with each new cycle, and dramatically escalating at our 67th recycle.
Cycle 1 Stars 1; Cycle 2 Stars 2; Cycle 3 Stars 4; Cycle 4 (8); Cycle 5 (16); Cycle 6 (32); Cycle 7 (64); Cycle 8 (128); Cycle 9 (256); Cycle 10 (512); Cycle 11 (1,024); Cycle 12 (2,048); 13 (4,096); Cycle 14 (8,192); Cycle 15 (16,384); Cycle 16 (32,768); Cycle 17 (65,536); Cycle 18 (131,072); Cycle 19 (262,144); Cycle 20 (524,288); Cycle 21 (1,048,576); Cycle 22 (2,097,152); Cycle 23 (4,194,304); Cycle 24 (8,388,608); Cycle 25 (16,777,216); Cycle 26 (33,554,432); Cycle 27 (67,108,864); Cycle 28 (134,217,728); Cycle 29 (268,435,45); Cycle 30 (536,870,912); Cycle 31 (1,073,741,824); Cycle32 (2,147,483,643); Cycle33 (4,294,967,296); Cycle 34 (8,589,934,592); Cycle35 (17,179,869,184); Cycle 36 (34,359,738,368); Cycle 37 (68,719,476,736); Cycle 38 (137,438,953,472); Cycle 39 (274,877,906,944); Cycle 40 (549,755,813,888); Cycle 41 (1,099,511,627,776); Cycle 42(2,199,023,255,552); Cycle 43(4,398,046,511,104)
Cycle 44 – Stars 8,796,093,022,208
Cycle 45 – Stars 17,592,186,044,416
Cycle 46 – Stars 35,184,372,088,832
Cycle 47 – Stars 70,368,744,177,664
Cycle 48 – Stars 140,737,488,355,328
Cycle 49 – Stars 281,474,976,710,656
Cycle 50 – Stars 562,949,953,421,312
Cycle 51 – Stars 1,125,899,906,842,624
Cycle 52 – Stars 2,251,799,813,685,248
Cycle 53 – Stars 4,503,599,627,370,496
Cycle 54 – Stars 9,007,199,254,740,992
Cycle 55 – Stars 18,014,398,509,481,984
Cycle 56 – Stars 36,028,797,018,963,968

Cycle 57 – Stars 72,057,594,037,927,936
Cycle 58 – Stars 144,115,188,075,855,872
Cycle 59 – Stars 288,230,376,151,711,744
Cycle 60 – Stars 576,460,752,303,423,488
Cycle 61 – Stars 1,152,921,504,606,846,976
Cycle 62 Stars 2,305,843,099,213,693,952
Cycle 63 Stars 4,611,686,018,427,387,904
Cycle 64 Stars 9,223,372,036,854,775,808
Cycle 65 Stars 18,446,744,073,709,551,616
Cycle 66 Stars 36,893,488,147,419,103,232
Cycle 67 Stars 73,786,976,294,838,206,464
(Over 73 quintillion stars)

Astronomers believe there are probably an even higher number of stars; somewhere over 1 sextillion (21 zeros) – which would make our universe even older. For the purposes of this book, 73 quintillion stars is referenced in our present 67th cycle of the star and solar systems of our cosmos.

Proclamation: Our universe as calculated by Trillion Theory at one trillion years of age and is over 67 times older than the 13.7 billion year Big Bang estimate. Our universe is celebrating its trillionth birthday.

Variance age of stars, solar systems and galaxies.

Each of the past 67 cycles of the contents of our universe lasted approximately 15 billion years. There is no exact cut off between one cycle and the next simply because many star lives overlapped into the next cycle. In our cosmos of planets, moons, stars and galaxies, there are threads from one cycle to the next. This occurs because of the variances in the life cycle of each star.

But in general, 15 billion is the normal length of the life cycle for a star. Therein, 15 billion years is also the normal length of the normal life cycle of a solar system as its contents recycle along when the star goes Supernova.

But, an ageless galaxy follows different rules because of the supermassive black hole at its hub which lives by a slightly different set of Universe Laws. Some galaxies are younger at 100 or 200 billion years, while some are 500-800 billion years old. Because different solar systems within a galaxy recycle on different time schedules or cues, an entire galaxy never has to totally recycle all of its solar systems at any one particular point in time. So, while all galaxies are much older than 15 billion years, their stars intermittently recycle each 15 billion years, just not all at once.

Implications of a recycling universe.

From all of this, draw two dramatic conclusions. Firstly, the universe had an inception. It started small, has grown, recycled, increased in population and expanded in size with each new cycle. Secondly, this universe will perpetuate to infinity; as each new cycle matter and energy will go back and forth like a ping pong ball, through eons of time.

Also, the universe will continue to draw new allotments of static light available to be spun into more new matter from the outer ocean of static light. Because of its method of construction and recycle, this universe is designed and programmed to perpetuate and grow onward into eternity as it has no possible end. It cannot be destroyed. Not by any means known of, and certainly not of its own accord.

Our universe is astronomically old, having seen a myriad of life times. From here on in, when you see a ray of light leaving our sun, you might see it in a different way. Not just as a ray of light leaving the furnace of the sun, but rather a ray of light caught in a recycle which has a long ancient past and history stemming back some trillion years.

As a ray of light leaves our sun, it passes by an endless unnoticeable graveyard where there once existed spheres of previous cycles. In a historical sense, it is unfortunate that we cannot see this old graveyard as evidence of those ancient dead spheres. Our universe's long history has been erased by recycles so complete that the remnants of the ancient past are nowhere to be found. Each old cycle is lost forever, while each new cycle appears uniquely individualistic, giving the optical illusion that it is the initial and solitary cycle.

The brightness of a Supernova is hitting us over the head to get our attention. Such a bright explosion is saying, 'Look here! See this exploding star; see the obliteration of a solar system; see the after effects as new naked black holes fight for light to spin as they begin birthing a new solar system.

Trillion Theory ages our universe at a trillion years represented by the 67 bars (each bar equals one 15 billion year time frame)

Each 15 billion years is one cycle. (A cycle can have some overlap with a preceding or following cycle). Take our solar system as an example, we are about half the way through our present 15 billion year cycle of our sun and after our sun dies or goes supernova, another 15 billion year solar system will grow from the surviving black holes which are at the core of every sphere in our solar system. However, any other next door neighbor solar system in our Milky Way Galaxy might be on a totally different recycling schedule.

Therein, ties always exist and overlap between one 15 billion year cycle, its previous cycle, and the next cycle.

Giving our universe a deserving name - Recycliun.

We could call our universe 'old indestructible.' A better name also describes operations. I like to call our universe 'Recycliun' composed of key letters taken from three key words, namely: Recycling (take Recyc) meaning that our universe recycles; from the word light (take li) indicating that the substance recycled is light; and word universe (take the un) stating that our universe is where the light is recycled. Therein, you have Recyc-li-un. Its action and duration is Relunoty (**Re**cycling **li**ght **un**iverse **o**ne **t**rillion **y**ears).

Trillion Theory shows how Recycliun operates.

It's actually quite ingenious how our Recycliun universe grew to its present 73 quintillion star count via its own super recycling methodology. Recycliun has recycled many times over. Alive and growing, doubling in size with each 15 billion year recycle, becoming more gigantic and expansive. Ever since its origin, Recycliun has witnessed continual moon, planet, star, and galaxy population growth with a doubling in size from one 15 billion year cycle to the next.

Our present 15 billion year cycle of our universe is but the latest in our universe's lengthy history. Unfortunately, the past recycles were so complete that all evidence of the near agelessness of our universe has perfectly recycled away.

The reason is that matter and energy recycle back and forth, leaving nothing behind as identifiable evidence. Since neither matter nor energy can ever be destroyed, they have the endless ability to recycle back and forth; matter back to energy; energy back into matter, over and over to eternity.

Where old stars go to die.

Where do old stars go when they die? Do they simply spread their material like star dust and never exist again. Why don't we see any long term graveyards piled full of old dead stars, similar to what we see when trees die in a forest?

No, that doesn't happen, because stars never totally die. When they go Supernova and seem to die, they cast away all their material, yet the minute (now naked) black hole at their core survives, ready to spin light into new matter to form a new body sphere around the black hole. This re-building might take a year if a supply of light is plentiful, or if not hundreds of years. Once again the black hole fills up and cloaks itself inside of a new sphere. Therein, any evidence that the old star ever existed is lost forever.

Replication of a black hole follows a Supernova.

While the star is gone, its black hole core remains intact. That black hole is small compared to the star it survived. The star, going Supernova, exploded emitting and unraveling back to light all the atoms which had comprised its body.

There are two indestructible materials in our universe which can never be destroyed, namely light and black holes. For the indestructible surviving black hole at the core, the extreme action of the death of the star forces the black hole core into a splitting action thereby doubling the black hole from one to two in its replication reproductive cycle. Hence, this splitting process produces two naked black holes to replace the one. There are now twice as many naked black holes prepared to grow into twice as many new spheres.

For example: In a solar system with one sun, 10 planets and 100 moons, if its sun went Supernova and wiped out each and every one of the planets and moons, the black holes at the core of every sphere would survive and reproduce into double the number of black holes ready to fight for light and build a new solar system.

This reproduction system for stars/planets/moons is a whole new discovery of Trillion Theory. This replication is the reason that our universe is so old and still growing.

CHAPTER 6
LIGHT FIRST THEORY

Which came first, a star or light? Everyone, following their eye would say that's easy, we see our sun (a star) and then we see light coming from that star. So, the star was there first and emits light. But, Trillion Theory differs, stating that before there were stars or space in our universe, there was only light. It was around before the stars ever formed. In Trillion Theory this is known as the Light First Theory.

This light was endless as a vast infinite ocean of light. So, before galaxies, solar systems, and spheres formed in our universe, first there was only a never-ending static ocean of light. Even the frontier of space was yet absent. This original light was frozen-like, stacked like lumber, as a non-moving static ocean of light, ready for use. Note: In 2014, physicists froze a light ray for the first time, slowing its speed to zero.

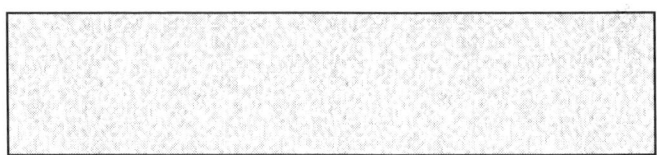

Before galaxies, solar systems, and spheres formed in our universe, first there was only a never-ending static ocean of light. Even space was yet absent. This original light was frozen-like, as a non-moving static ocean of light, ready for use.

This static non-moving ocean of light contained and held, in every possible direction, tightly packed strands of light. All material properties necessary for sphere-building were waiting and contained within each strand of light in the stack; namely the light spectrum, heat, weight, and mass.

Then, the first naked black hole (the engine for building) was introduced into the physical universe's static ocean of light a trillion years ago. Envision a black hole interjected into the ocean of light, a naked black hole set to consume light from the ocean of light. The commencement of the building of the very first sphere of our universe had begun by that black hole. Immediately, that naked spinning black hole began to loosen and break light strands away from the ocean of light. The naked black hole quickly consumed the broken away light and spun those light rays into atoms of matter inside and around the black hole. Huge amounts of the strands of light were consumed from the ocean of light by the black hole. Soon, a round-bodied sphere began to build around the black hole. The first planet of our cosmos was forming. It had weight and mass which light supplied, and spun on its axis around the black hole inside its core.

Back at the ocean of light, where the huge amounts of light had been taken from, the ocean of light now saw a vacant area. This vacant space surrounded the planet, but this space now lacked light, heat, weight, and mass. This space occupied the entire large vacated area inside of the still prominent static ocean of light. Space became the weightless empty black void which light used to occupy.

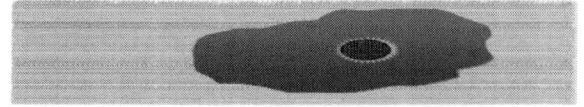

Within the bounds of the static ocean of light, the first naked black hole was interjected. Immediately, that spinning black hole loosened, broke away, and consumed light from the ocean. Those light rays were quickly spun into atoms of matter inside and around the black hole forming a sphere. That sphere had weight and mass which light had supplied. And, black space became the weightless empty area which light used to occupy.

To capture still more light, the black hole worked its way continually deeper into the ocean, consuming tons of more new light. As the black hole gnawed away at the ocean wall of light, each strand of light which it broke free attempted to move from zero to the speed of light as quickly as possible. However, that freedom only lasted a second for the ray of light as the black hole captured and spun it into matter.

Eventually, the black hole ate its fill of light, forming an entire heavy sphere around itself. This sphere of formed matter (atoms spun from light) spun on the axis of the black hole and continued to project a gravitational pull as the interior black hole still attempted to pull more light in towards itself. But now full, the black hole could only attract any free light to its surface, unable to spin any more light into matter. The black hole went from the phase of spinning of light into matter, to a holding pattern for billions of years attempting to maintain control of the tons of light which it had spun into uncountable atoms of matter.

On the other hand, the spun light would now be jailed as matter for billions of years, waiting to someday escape and be able to travel through space as free running straight line light. That someday would be when the black hole's holding abilities tired, thereby the surface of the sphere loosened enough to become that of a sun, and atoms of matter could unspin and escape from the sun as free traveling light.

Indestructible light would have recycled back to free traveling straight line light after several billions of years of entrapment inside an atom of matter. Light would have finally escaped the clutches of the black hole.

In summary, light preceded the star, and outlived it too. The catalyst naked black hole ate from the ocean of light, forming atoms of spun light into a sphere, leaving a totally vacant area called space between itself and the light ocean.

Trillion Theory and Light First Theory provide a new 'big picture' look at our cosmos.

There is a need to really rethink our universe. We must now view light, matter, and space somewhat differently. In Big Bang, we were taught that matter exploded cross space. Whereas in Trillion Theory, material light was there first as a static ocean of light and naked black holes accessed this light to build the spheres of our universe; space is simply a left-behind empty highway of this building process.

So, in Trillion Theory, at the origin of our cosmos, there existed only light in every possible direction; this three dimensional ocean of static, non-moving, light as complete and as incredibly voluminous as the water filling an ocean.

These new concepts place a whole new origin to our universe. New theories in this book illustrate how our universe began very small by forming a single tiny solar system a trillion years ago. And then exhibit how that solar system multiplied and grew into today's gigantic universe.

The catalyst to begin the process of spinning light into matter occurred via the introduction of a black hole (sphere builder) which was cast into the static ocean of light. That black hole spun away at the static light, breaking some free. When, the light broke free it went from motionless light to free traveling light. The black hole then captured, bent and spun that light into matter around its black hole core.

The amount of light required to spin the atoms and body of matter around a sphere is immense. The formula $E=mc^2$ calculates the huge ratio amount of light energy which goes into a relatively small amount of matter formed.

'What comes out is what originally went in.'

An atomic explosion releases a tremendous amount of light energy from an atom. Therein, Trillion Theory says that, 'What comes out is what originally went in.'

Only one material.

For the next moments, the reader is requested to think more like an inventor of a universe. How would you build a universe? Of course, incredible science would be required. But, simplicity might also be an important factor. And, the cleverness of disguising building methods might certainly be a strategic ploy. Building-in a feature such as perpetual spin would be one way to give the universe inner movement and recycling abilities, so that it could grow of its own accord. Two of the greatest searches would be for an all-encompassing building material and then an engine to make the entire system function and operate.

Trillion Theory states that the building material deployed in the building of our universe was 'light,' which is a most astonishing material possessing a vast array of absolutely incredible properties. Trillion Theory claims that 'light' was the sole material used in building our entire universe. Every star, planet, moon, solar system and galaxy was built from one material, namely light.

Our universe's design was quite the gargantuan task, by whoever did it. Incredibly this was accomplished through the deployment of just one basic indestructible material, namely a material we see each and every day, namely light.

Thus, Trillion Theory maintains its Light First Theory. Light existed in our universe before and ahead of any matter. Light First Theory maintains that all universe spheres came about as the result of light being spun into matter.

And the engine which spun light (the sole material of our universe) into matter was black holes - the sphere builders.

Properties of light in Trillion Theory.

Light is the most fantastic amazing material in our entire universe. Yet, this light seems to be extremely humble. It has taken many scientists thousands of years to unveil some of its hidden powers; and yet there is still more to discover in order to understand the total properties of light.

Light is extraordinarily versatile, able to act like a particle at one moment and then as a wave the next. Light can bend as we see evidence in a rainbow. And furthermore, Trillion Theory maintains that light can bend to such an extent that it can spin into matter. We don't see this happen around us because it takes the special gigantic rotational force and speed of a black hole to capture and spin light into matter.

Through space, light travels as a single ray in a straight line for billions of miles, speeding towards some unknown destination to make a delivery. Underneath, light carries a payload, a hidden tool box. Its motto: *Always be prepared.*

Trillion Theory properties credited to LIGHT:

♦ Light is thus far the fastest traveler in our universe. It can travel at a speed of 186,282 miles per second; 5.9 trillion miles in a single light year; one parsec in 3.26 light years.

♦ Light, traveling in a straight line at the speed of light, carries its own tool box: electromagnetic spectrum; heat; weight; time. Its motto: *Always be prepared.*

♦ Light, when it travels, it is always ready to supply its tool box to be spun into matter.

♦ Light's speed stays at its high constant unless slowed by a gravitational force. Light slows and bends when it meets with the gravitational pull of a black hole at the center of a moon, planet, star, or galaxy, or when passing through substances providing extra resistance.

♦ Light, when it slows, supplies its properties into the area. Light slows when it encounters the gravity of Earth and supplies light and heat. But, this particular light is not spun into matter as the black hole at the center of our planet Earth is too full to spin more light into matter.

- Light's speed slows even more when encountering the powerful gravitational pull and force of a naked black hole which attracts, bends, and spins light into matter.
- Light, when it spins, supplies even more of its properties into the area within and around the black hole which has spun it into matter. The spun light becomes the atoms of matter. Light supplies the atoms structure, weight, and interior movement. Light also brings to this matter the concept of time. (Time discussed in a later chapter).
- Light is thus the universal supplier, the building material, for making everything known as matter in our universe.
- Light is totally indestructible, thus the ultimate recycling material, recycling on to infinity. Straight line light can be attracted and spun into matter by a black hole and then at the end of billions of years in captivity, that light can escape an atom returning to straight line light.

Light meets gravity

Gravity is a pulling force which surrounds all the spheres, galaxies, and black holes in our universe. Certain spheres, acting as *Gravitational Stations*, operate differently than do others. (Gravity is explained in a later chapter).

Upon arrival at a gravitational station (moon, planet, star), light bends and also slows below its straight line speed of 186,282 miles per second. Light further slows when it meets an atmosphere where it can be absorbed, reflected, bent or refracted (rainbow colors ROYGBIV). Light further slows when passing through higher density: in a diamond, light slows to 77,500 mps. In all of these instances, light can only be forced into a bending stage. In actuality, the black hole at the center of the sphere supplies the gravitational pull. That gravitational pull goes from the black hole right through the entire matter of the sphere and extends out past the sphere into the surrounding space. Some of this light is absorbed by the planet and some is reflected away, back into space. The hard matter of the exterior of the planet prevents the already full black hole in the planet's interior from getting its hands (so-to-speak), in direct contact with the light and spinning it into matter.

Light traveling into a gravitational station, immediately opens its tool box and sets to work, either providing light for vision and heat. Type of work depends on the gravity station encountered. Light, 100 hundred percent of its time, is constantly 'doing' as a universal supplier. Light's other motto: *Always try to be in constant motion.* Light is the most absolute inexhaustible, indestructible building material.

Laws of Light when encountering the ferocious spinning speed of a naked black hole:

- Light slows more as it is grabbed by the black hole's strong ferocious spinning speed.
- Light bends to the extreme, coiling and spinning.
- Light spins forming the body of an atom.
- Light, once encased inside an atom, can spin inside the atom with varying lengths of strands.
- Light with varying lengths and thickness of strands form different atoms of elements.
- Light can have its head as an atom's nucleus and its tail for the electrons.
- Light has the property of weight in its tool box providing weight to the atom of matter.
- Light spun into atoms of matter initially fills only the interior of the naked black hole.

When light encounters a gravitational station known as a powerful naked spinning black hole, light is easily attracted, refracted, bent, spun, and assimilated into atoms. This light loses more speed as each action occurs.

Those initial assimilated atoms fill only the interior of the black hole. Once that is full, more atoms form the exterior body. After an entire sphere body forms around the core, the black hole can only attract, but not spin more light.

Laws of Light once interior black hole filled with matter:

- Light spun into atoms of matter at the interior of the black hole continues to spin as an atom and will be locked away in their interior jail for billions of years.
- Light spun into more atoms will form the body of matter surrounding the black hole. A small solid sphere is born.
- Light attracted by the gravitational pull from the interior black hole adds to the exterior sphere's surface forming it into a moon or planet. Rotational speed at the black hole's core is slowed by the added matter.
- Light turned into matter as an atom can continue to spin for extreme eons.
- Light trapped as matter always wants to escape.
- Light trapped as matter at the interior of the black hole's core becomes exhausted from its toil.
- Light trapped as matter at the interior of the core will begin to unravel first, with seemingly nowhere to go as it is trapped. This matter will heat up.
- Light is always successful in its escape from matter, even if it takes billions of years.

Light has a formula of escape: $MC^2=E$. The total amount of energy (E) released from a single atom of matter depends upon its mass (m) multiplied times the speed of light (c) squared (what comes out is what went in). That formula tells of the tremendous amount of light entrapped within a single atom of matter waiting for eons like some genie to escape.

Everywhere, these Laws of Light are relevant.

We were not around for the origin of our universe a trillion years ago. We weren't even around for the beginning of the present 67th cycle of our universe when our present planet and solar system formed billions of years ago. In neither case did we get to see what material was used to construct the many atoms of matter of basic elements. We missed out on that monumental manufacturing process.

Because of that, in our present day, we live on a planet where everything has already been spun into atoms of matter. All we do is take those already manufactured atoms and use them in thousands of ways. Man has learned via chemistry to mix and match these elements into compounds and even macromolecules (polymers) for plastics.

But, the main point here is that on Earth we missed out on seeing the initial creation stage where light energy was spun into atoms of matter. Now, we mostly see the locked stage (light locked up inside atoms).

However, on planet Earth we have been provided with some end-clues: sometimes we get to view a partial end phase such as lightning or a nuclear explosion when varying amounts of light escape atoms. And, every day we get to witness an end phase when light escapes from the sun. That light departs an unraveling helium atom on the surface of the sun and escapes back to freedom as straight line light. After that end phase, light will either continue to travel for millions of years through space, or meet up with a waiting black hole only to be captured and respun into matter once again (new creation phase).

The real magic was how science could have invented and designed the most incredible material (namely light with a host of properties) and powerful engine black holes (housed with specific powerful makeup to create mega spin), and combined the efforts of both light and black holes to work together sometimes as friends, or sometimes as foes, to make the most beautiful universe imaginable."

An analogy of light spinning into matter.

When it comes to the making of an atom of matter from light, a useful analogy is a grandmother knitting. Think of light as yarn. Notice the beautiful knitting as the lady makes many wonderful garments from her yarn. She uses different amounts of colorful yarn, various types of knots, a variety of patterns, and different tightness of loops, to create a warm woolly scarf, a toque, sweater, mitts, socks, or blankets. They all look different, but underneath they are all made from one material, for they are all just yarn. She can even take different garments and sew them together (molecules).

Then in reverse, the grandmother can take every one of the knitted garments and unravel them back to their original yarn material. Thus, regardless of the disguise seen in the garments, they are all just yarn.

Therein, following on that theme, when you unravel all atoms of matter you discover tons of escaping light (along with the properties of light). To take that thought further, when you unravel all the atoms of a moon or planet or star, you have an absolute incredible amount of light energy released from captivity. Formula $MC^2 = E$ comes into play. The amount of energy volume which escapes back to straight line light would be a staggering to a zillionth zeroes. And going back to the creation phase, the amount of light spun into matter to originally make a planet was staggering.

The author realizes that it may be difficult for the reader to accept that all matter is made from one and only one substance, namely light, when your eye sees such an array.

But, light has unique properties which allow it to be spun into a whole number of different atoms of specific types of elements. There exists on planet Earth over 100 discovered pure elements of matter, and every one of these elements are specific type atoms. There may even be other elements on other planets or in other solar systems or other galaxies which we have yet to discover. At any rate, every one of these 118 known elements is a pure atom with its own specific design with a specific tightly packed nucleus (with protons and neutrons in that nucleus) and a specific number of electrons in orbits circling that nucleus. Each element atom is unique from every other element. Many combine as molecules to add to our world's phenomenal complexity. Each of these pure elements are different, yet the same. They all are atoms with nuclei and electrons, all supplied by the same material, light. They differ only as each unique element has a different atomic arrangement.

Far-fetched.

As the reader of Trillion Theory, at this stage you may have definite reservations as to how these things in our universe could be; how light could be the one and only material making up everything, or how black holes spun light into matter and built the spheres, solar systems, and galaxies of our universe. You might even say, "Too far-fetched and too improbable, that our cosmos grew larger from its meager beginnings a trillion years ago to its present colossal size and sphere population."

Author's reply.

Any more far-fetched than the Big Bang? And Big Bang fails and does nothing to explain thousands of things throughout our universe. Whereas, Trillion Theory attacks the mysteries of our universe head-on, searching for the truth. Please be patient and wait for the entire Trillion Theory, start to finish, before passing your final judgment.

CHAPTER 7
BLACK HOLES
ENGINES MAKE MATTER
(LIGHT INTO MATTER)

How black holes accomplished the monumental task of building spheres in our universe.

This was done (and is presently being done) by black holes spinning light into matter on such a large scale per black hole that an entire bodily sphere such as a moon or planet forms around the spinning black hole.

How this began.

Remember the *Light First Theory*, where light is credited as the singular building material of our universe, and where it preceded everything else in our universe; a waiting ocean of light, zeptillions of strands of piled light in all directions, available as an ever-supply of energy for our cosmos.

Then, at the origin of our universe a trillion years ago, the catalyst engine to begin the process of spinning light into matter occurred with the introduction and casting of the first black hole into the static ocean of light. This singular event began the entire growth process of our universe; a universe which has been growing ever since.

A singular Black Hole appears in the ocean of light at the origin of our universe. It contained incredible spinning power.

With the introduction of the one initial naked black hole into the ocean of light, it did the duty it was programmed for. Deploying incredible spinning speed, the naked black hole broke light free from the static ocean of light, attracted it, pulled it in, and then spun the light into matter.

Had we been there to witness that memorable scene, we would have seen the blackness of the core of the black hole with the brilliance of light rushing into it. The black hole would have used its spinning power to crash against the static ocean of light, breaking away chunks of frozen light, freeing them to release and begin moving as free straight line light, but not for long. Those free rays of light would have been quickly and methodically devoured and spun into atoms of matter inside the belly of the black hole.

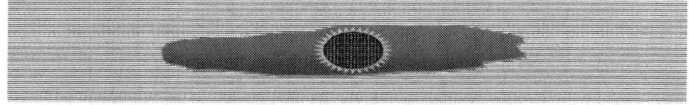

As the singular black hole ate from the static ocean of Light, it spun light into matter around itself. Light brought along its tool chest of properties including heat and weight. Empty weightless linear space was left behind as gapes in the static ocean of light.

As it spun more light into matter, the black hole filled its belly full of matter first and then begun to increase the size of its surrounding ball of matter. This was the first planetary sphere occupant of our universe. Around that ball of matter, between the ball and the ocean of light, space formed as the empty spacious byproduct left behind when huge amounts of light consumed by the black hole exited that area.

This empty space formed around the black hole was linear in shape, formed as an extension of the gravitational force of the axial spin of the black hole; meaning that the black hole ate most of its light from an extended equatorial plane.

A black hole is sphere shaped. When it spins on its north-south axis the gravitational pull created holds matter on the surface. But, outside the black hole the gravitational pull is on a linear horizontal plane which extends out past the equator where it can hold other spheres in orbit. (Like Saturn and its rings).

This linear result comes about because of the spinning action of the black hole on its axis. The gravitational pull caused by the spinning black hole projects outwards from the black hole on a plane horizontal to its equator. This linear flat effect is demonstratively seen in the orbit of the planets in our solar system around our sun. This is also seen in the disc-like shape of the rings spinning around the planet Saturn; and with the disc shape is a spiral galaxy; and with the shape of our entire linear, not spherical, universe.

Once that linear oval shaped plane of space built around the extended equator of the black hole, the black hole had to move through space to keep close to the ocean of light.

One might ask why more light wouldn't have just flowed from the ocean of light into the now empty space. The answer is that the light ocean is static, comparable to frozen. To gain light from the ocean, the black hole must loosen more static stands of light, free-flowing them.

As this process continued on, the black hole grew more matter around itself, growing into a sphere. The width of space around the black hole grew ever larger as more light was taken from the ocean. Thus, the size of our universe began to grow with the advent of a growing planet and the expanding boundaries of space. The static frozen-like ocean of light still occupied the outer perimeter, while inner space became a larger linear-shaped empty void.

Thereafter, as more and more black holes ate from the ocean of light, many more spheres were built congregating into solar systems. The congregation into solar systems was due to the gravitational forces supplied by the black holes inside of the spheres. These black holes used the force of their gravity to hold other smaller spheres in orbit.

Till Trillion Theory, no one knew.

Till Trillion Theory, black holes were never thought of as having a definite duty in our universe. Till now, No one knew their true purpose as builders of the spheres of our cosmos.

Till Trillion Theory, a black hole was never thought to be at the center of every moon, planet and star.

Till Trillion Theory, no one had ever surmised that light eventually escapes from a black hole even after being trapped inside for billions of years.

Till Trillion Theory, no one had theorized how the eventual escape by light from a black hole is part of the recycling process of the spheres in our universe.

Till Trillion Theory, no one had theorized that black holes are actually alive in their own way; with an animalistic-like appetite to devour light and hold it tight; to become the largest and strongest and capable of dominating smaller spheres; able to survive and multiply through the ages; to be the dominant engines and builders of our cosmos.

Till Trillion Theory, no one had theorized that there was a living-type of entity at the core of planet Earth and every other sphere (moon, planet, and sun) in our universe. This entity does not die when our Earth ends. Rather this entity survives as a naked black hole, splits into two during fission reproduction such that two new naked black holes begin the next 15 billion year cycle of a new solar system. Black hole survival and replication, and an oceanic supply of light are the key reasons our cosmos grew to over 73 quintillion stars.

Albert Einstein laid the theoretical foundation for the existence of black holes by predicting that light would bend when nearing a sphere's gravity. However, no one had ever surmised that black holes spin light into matter; or, surmised the existence of a black hole living for billions of years at the core of every sphere (moon, planet, star) or at a galaxy hub.

On the other hand, only in short periods of time do astronomers get to telescope a naked black hole wherein it has shed its outer sphere, and is in the process of attracting and devouring new light in its attempt to fill its belly and then its exterior with light spun into matter so that the black hole can reside at the core of yet another sphere.

Because most of their time is spent hidden away, black holes were never fully understood. Humans had no original inkling, till now, that black holes were related to the creation of spheres and to the recycling process of our universe. Rather, black holes were simply thought to be rogues in space where light disappeared into a pit which possessed phenomenal gravitational forces, never to return.

You see, the monumental discovery of black holes by astronomers was just a first step in uncovering their secrets: Astronomers have incorrectly thought of them as simply rogue-like invaders of space, looking to devour a star, for no good apparent reason. Now, Trillion Theory gives black holes a whole new importance, making them prevalent throughout our universe. They are instrumental centers of birth and rebirth of every universe sphere.

Just like E=mc², where both energy and matter cannot be destroyed but only interchanged, black holes are master survivor recyclers. They can never be destroyed even by the power of a Supernova. Try to destroy a black hole and it simply splits and reproduces into two new black holes to once again begin the pursuit of building a new sphere.

Over the last trillion years, black holes have been *sphere factories* hard at work building and recycling the spheres of our universe. A fundamental component of Trillion Theory is how our universe originated. Today's enormous cosmos is a product of black holes being recycling sphere factories.

Initial Conclusions.

Black holes were and are the beginning mechanisms that operate like an engine (a spinning rotor) spinning light into matter around itself.

This tremendous scientific-type invention of light and black holes is absolutely extraordinary. The 'tool box' within light, and the powerful spinning mechanism of engine-like black holes, were both inventions made to last and endure forever. It was pure genius that the origin of this universe came about from light being devoured by a black hole. It is also pure genius that this universe grew from the interaction of light and black holes over a trillion year history.

A big question.

From where did the original static ocean of light strands come from and also the initial black hole?

That is a huge question, to be sure. To know for certain one would need to go back to the scientist(s), or Artisan(s), or creator(s) of this universe (if such an entity or entities exists), knock on the door, and make that query. Whatever the answers might be, incredible scientific knowledge (far beyond ours) was deployed in the inventions of both light and black holes. Also, there was a clever strategy deployed in the methods of operation in our physical universe. As to who or what did the creating or supplied the materials, I've always loved a mystery. I believe that whoever has been cleverly egging us on, tossing us tidbits and saying "look here! rainbows; nuclear explosions; supernovae are clues."

The various masks of a black hole.

A black hole gives different looks depending upon the amount of light which it has or hasn't spun into matter around itself. Naked black holes appear just as the words describe, naked and empty as they are just beginning their search and spinning of light into matter. This is the best time in the lifecycle of a black hole to view its active components.

However, once a black hole has spun sufficient matter, it can hide inside the core of the sphere of matter it has spun. These cloaked black holes can be hiding behind a number of masks or disguises, as a moon, a planet, a star, or at the central hub of a bright cluster of a galaxy. Black holes give different looks, dependent upon the stage in their cycle:

A black hole has major phases as it does cycling.

In our Recyliun Universe, recycling of spheres follows the same 5 major sequence phases for one cycle of a black hole to a sphere, and then back to a black hole. These 5 phases normally add to 15 billion years on average.

During one fifteen billion year cycle a black hole goes through phases. Two examples: Our present Earth is in a mid-cycle planetary phase, so the black hole at the core of our planet is masked behind the hard exterior coating of our planet. While, our sun is in a late-cycle star phase, so the black hole at its core is masked by the fiery exterior. Note: A black hole at the center any sphere could have a shortened cycle if the sun of its solar system goes Supernova.

At the beginning of each new cycle, a black hole always starts out with a naked phase. But as it consumes plentiful light which it spins into matter, the black hole can mask and hide itself inside of a sphere such that we see its outer appearance in the stature of a moon or planet; or as a sun with a fully-fledged solar system under its control; or even as a gigantic black hole at the center of an entire galaxy.

Phase One – Naked Acquisition and Creation of Matter

We see a black hole spinning light into matter.

There is a supply of light rays, either from the static ocean of light, or as free straight line light traveling through space.

Phase one, the naked acquisition of light and creation of matter, always begins with a dark empty naked black hole beginning to spin light into matter. Naked meaning that the black hole is empty and has not yet spun light into matter around itself. This is the type of naked black hole introduced into the static ocean of light at the origin of the universe; also the same type of naked black hole left behind when a star goes Supernova exploding away all its trapped light.

The first assignment of any naked black hole is to attract, capture, spin, and entrap straight line light into extremely tight dense matter. In this CREATION OF MATTER phase, the hole appears as black as black can be. However, brightness can occur when tons of light is gobbled up and a stream of light is seen gushing inwards towards the black hole.

Phase one takes a short time frame. 'Blink and you might miss it.' Duration depends upon the amount of light directly available for the black hole to spin into matter. If the black hole is close in proximity to the static ocean of light, it can eat and fill up quickly within just years. If far from the ocean of light, and having to fight against other naked black holes for light, the filling process might take hundreds of years.

Creation of Matter phase:
black hole spins light into matter in and around itself.

The black hole has more than just the ability to spin light into matter in its core. It has the power to use its tentacles to continue to spin light into matter and add that matter in layers around itself. These layers build up and become the outer surface of a sphere forming around the hole.

Phase Two – Lockup Control

Lockup occurs once the black hole becomes full. This total lockup control phase can last 5-10 billion years. We see a small sphere as a moon or planet.

The lockup control phase follows the initial spinning of light into matter phase. The teen-aged black hole has spun light into matter so that the size of the ball formed around itself increased dramatically in size. Once a black hole has spun sufficient matter it no longer appears as a naked black hole. Now, the black hole is disguised and cloaked (hidden) away inside the core and under the surface matter of the moon or planet which it built.

As it adds more matter around itself, the black hole may have spun enough matter to be a moon or planet. The size of this sphere is directly related to the size of the black hole at its center and the amount of light is has spun into matter.

Once this maxed-out size is achieved and the black hole has finally spun as much matter as it can for the size of its black hole, duties for the black hole shift to lockup and control. All effort now switches to the new task of control (LOCKUP), trying to prevent matter from unraveling to light.

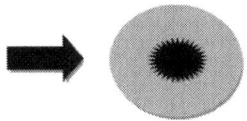

Black hole lockups up light as atoms of matter around itself.

The many layers of matter comprising the main body of the moon or planet are held in place by the strong gravity hold of the black hole's helix, its core compartments, and tentacle filaments. Everything is held in place by the force created from the power spin of the black hole. This gravity extends beyond the surface of the sphere to attract and curve more light inward towards the sphere, and extends far enough outwards to hold any other smaller spheres in orbit.

While the black hole still wants to spin more light into matter (in its desire to grow ever larger) that task becomes ever more difficult as it fills up. Each new iota of matter which the black hole adds to its exterior surface blocks its ability to be directly in contact with and spin new light into matter. It can still attract passing light to its surface, but the more the black hole thickens its surface, the less ability it has to spin by-passing light into new matter.

Eventually, the spinning speed of this fuller black hole begins to decelerate. The rate of spin at the Spiral Helix of the black hole decreases as it becomes sluggish from all the matter filling all its compartments and its surface. Now, this slower speed really states that this is a planet/moon and no longer just a black hole. Light, while it can still be attracted, is simply absorbed by the surface, or even reflected.

The black hole now switches its focus from spinning more light to the task of control over the zillions of atoms which it has already spun into matter; trying to prevent the spun atoms of matter from unraveling back to straight line light.

Phase three – Loosening and some Escape

Lockup tries to continue even while LOOSENING begins (we see a planet with a more active surface). Once this loosening gets more rampant, ESCAPE commences. This phase can occur as early as 5 billion years and last 10 billion years.

After years of first spinning light into matter, followed by billions of years of controlling and holding that matter in place, the moon/planet begins to weaken. Because the black hole at the core of the moon or planet always overate, (always over-consumed light), the day comes when it must pay for having to maintain total control over too large an amount of light spun into matter.

Let's examine how a black hole tries to maintain control over the matter which it had spun from light, and how a loss of some control results in the beginning of the loosening phase of the contents of a moon or planet.

When the black hole fills from its eating, it still attempts to draw more light to its surface and tries to devour that light and spin it into matter, but it can't because the matter on its surface blocks the way. An example would be Earth.

Inside Earth, our black hole resident has strong control over surface atoms (we don't float away into space), but with this strong attempt to hold matter on the surface and keep our moon in orbit, and attract light to the surface, our black hole has lost control over atoms deep inside its core. Those oldest spun atoms are the first to begin the process of LOOSENING. The nuclei of these oldest atoms now begin to loosen as they spin less tightly. They begin to expand in size, and unspin as they attempt their ESCAPE. However, for loosened atoms there is nowhere to go as they are buried deep below miles of matter. Yet, this loosening does create heat inside the core of the black hole as the core fires up, and this heat causes expansion.

The planet around the black hole grows larger. The black hole at the center is now small in proportion (100's of times smaller than the matter it has spun into the planet's size).

We can only imagine the turmoil for the black hole as a hot fire expands begins deep within its belly, commencing its billions of years of losing control. That heat and fire in the black hole's internal belly expands and places tremendous pressure on the entire body and structure of the sphere.

Fissures form in the sphere's structure as lava created from the internal fire fiercely expands outwards searching for an escape route towards the surface of the sphere. When those fissures are blocked by solid matter, something has to give as the surface structures undergo dramatic change.

The internal pressure can force areas which are under or at the surface to buckle projecting large masses of matter upwards (mountains form). The planet mushrooms from the pressure. The surface itself is forced to expand (continental drift occurs). Eventually, lava breaks through as a historical first volcano (Pimple Burst) explodes gigantically to relieve the pressure. Volcanic activity for the sphere will peak.

This lava which reaches the surface can quickly cool and form rock, or some of the atoms totally escape billions of years of jail and escape their atom making their way back to free straight line light as they depart our planet.

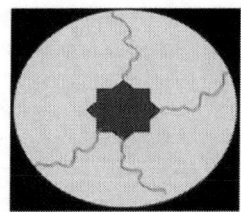

The oldest spun matter at the core of the black hole releases first and unspins to form a fire which searches for a way out through fissures to the surface to form the lava of volcanoes. Once this process begins it is irreversible. It may take billions of years, but in the end a fireball star is the final result.

If the planet survives long enough, lava eventually engulfs the entire surface transforming the sphere into a fireball star.

This loosening phase of a moon or planet could include 100's of sub-phases over 5-10 billion years. Looking at our solar system, although all spheres are similar in age, their state (solid, or gaseous, or plasma) is dependent upon the size of their black, how much that black hole overate, and how well that black hole has been able to maintain control.

Earth (a solid surface) has a hot interior where the escape of atoms to the surface has been going on for millions of years. It will be many more billions of years before Earth can loosen more and have a more gaseous liquid surface.

Jupiter, on the other hand, the largest planet in our solar system, really saw its black hole overeat during construction, and its interior is extremely hot. This heat has turned the surface into a more gaseous liquid state. Trillion Theory says that had it not been for our sun, Jupiter would have been the top candidate to become the sun of our solar system.

Sun has the largest core black hole in our solar system. It ate by far the most light when our solar system was forming. Sun was glutinous as to how it pulled in light and spun it into matter. Our sun, with its size dominance, formed the biggest body and also gained central status as the most prominent gravitational force holding all other lesser sized spheres (planets and moons) in sun's orbits. Our sun was also the first of all the spheres in our solar system to loosen control over its matter. Fire started early in its belly, flowing uncontrolled to the surface, a surface which is now totally gaseous/liquid known as plasma. On this plasma surface, atoms now easily unspin and release light free to travel through space, or until meeting another gravity station.

Phase four – Supernova and Obliteration

Occurs at age 10-15 billion years towards the end of the star's life. For a glorious sun, escape becomes rampant, as the sun has solar flares and emits tons of light.

Phase four deals with the final phase in the lifecycle of a sun. Note: Very few planets or moons ever reach star status, as most of these lesser spheres are destroyed when a sun goes Supernova and obliterates its solar system.

In phase four, the sun loses control of its contents. The sun of our solar system is in the mid-stages of phase four. It has expanded in size, growing ever larger. The fatigued star's contents loosen and expand; each atom becoming larger, thereby expanding the suns overall diameter. The sun now emits light from its surface at a hectic pace, freeing atoms back to straight line light; a light which then warms planets and moons in the sun's solar system.

Hot atoms are now prevalent from the black hole at the sun's core right to its entire plasma surface, where solar flares leap out from the overall intensity of the absolutely extreme heat. A fireball is the result.

In the end, the star known as our sun, will expand and eventually die exhausting all its fuel. When our sun expands gigantically, dies and goes SUPERNOVA, the expanse of this explosion will be far reaching. The Supernova will destroy some or all the planets and moons within our solar system, depending upon a sphere's proximity distance from our sun. Prolific carnage of a solar system by a sun - OBLITERATION.

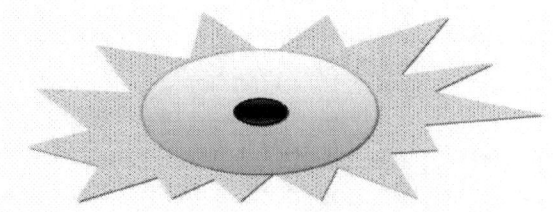

At the point of total fatigue, the black hole at the core of the sun loses control of all its spun matter as the sun's contents explode going Supernova as the sun dies.

Imagine the blast from an uncountable number of atomic explosions occurring simultaneously when any sun goes Supernova. The heat from all the instantaneously escaping light melts all of the spheres (planets and moons) in that sun's solar system. This Obliteration destroys all the surfaces, leaving naked black holes from the core of every sphere as the only survivors. Additionally, each sphere in that solar system will experience a flash Supernova of its own melted away body, thereby increasing Supernova's overall power.

After the Supernova of a star and obliteration of moons and planets in that solar system, tons and tons of pent-up light escapes from the matter which had been locked up on and within those spheres. That freed light goes back to straight line light, free to travel through space, until called upon to spin back into matter.

However, much of that happy freed light only experiences temporary freedom, as the now naked black holes which survived the Supernovae commence their battle against one another to spin that light into new atoms of matter for the building of the bodies of their new spheres. The creation and the building of a new solar system begins.

Phase five – Reproduction (Replication)

After Supernova, a new start up phase occurs. This takes only less than a hundred years. We see a graveyard of naked black holes which survived Supernova. For every old black hole we now see two new naked black holes.

When the star gigantically expanded to go Supernova, all of its contents of spun matter changed back to light which escaped from its surface as straight line light. Surviving was the now empty naked black hole which was at the sun's core. This pliable elastic black hole can never be destroyed.

At this point, this now abandoned black hole undergoes an amazing transformation. A black hole can now regain its prolific spinning speed. With the terrific unraveling force of a Supernova, the now exposed black hole regains its prolific spinning speed, as its core spins furiously in the opposite direction (for every action there is an equal and opposite reaction – Newton). From the violence of the Supernova, and a ferocious counterspin, this reverse force splits the helix of the black hole in two, splitting the entire core of the black hole as well. The two split parts of the black hole recoil and shove hard against each other. Thus, two new duplicated black holes replace the original one; REPRODUCTION, now two black holes instead of one. Replication is the end-phase of the old cycle; now begins the next creation phase.

Immediately these two new black holes go back to their virgin instincts and begin competing to devour light. Their first use of the new light they spin into matter is to fill their gut and then to increase their size by the doubling of their compartments via fission. Soon, the two new black holes are as large in size as was the original black hole.

During Obliteration, the destroyed moons and planets in the solar system explode near instantly when their central sun's Supernova destroys them. While their emancipated bodies have returned to straight line light, each black hole remnant which was at the center core of each destroyed sphere survives the obliteration, (black holes can never be destroyed). Each black hole axis splits in two from the force of Obliteration. The end reproduction result is a twinning of all black holes in the region, doubling the numbers heading into the next cycle for that area. Right after the Obliteration, the area around all the new naked black holes becomes a war zone as all the new holes cannibalistically fight over and snap stands of light to spin that light into matter.

Every moon, planet, and sun destroyed by the Supernova has a split occur at the site of its abandoned black hole.

Example: If the entire destroyed solar system had one sun with 6 planets, and 16 moons = 23 spheres in total, then the after effects of the Supernova would see those 23 spheres now doubled into 46 new naked black holes.

The black hole which was at the center of a destroyed sphere splits to form two new black holes.

How to view our universe's phases

Unfortunately, because of short human lives relative to universe time, our view is limited. Of the phases, we see mostly the lockup stage on planet Earth. However, powerful telescopes can show us Supernova and Creation phases.

For a human living in a planetary setting to see the precise internal creation process within a black hole at long distances is virtually impossible, even with the advantage of powerful telescopes. However, it might be possible with evermore powerful space telescopes to catch black holes in the process of creating the moons and planets of a far away new solar system. Such a view would prove Trillion Theory.

Not every black hole which spins a sphere around itself sees that sphere grow into a sun, few evolve that far.

During Supernova, the explosion of the sun can have far reaching Obliteration effects on the rest of the spheres of that solar system. During Obliteration, all planets and moons are torched by the exploding sun such that all these lesser the sun-size spheres face their own instant supernova.

After Obliteration occurs, naked split black holes (double the number after Reproduction) are the graveyard survivors of all the supernovae which occurred within that particular solar system. That graveyard will only be visible for a short period as all naked black holes in the graveyard will begin their battle to spin light into matter. One new large solar system or two adjacent solar systems can be the end result.

Summary of light and black holes.

Take away light and black holes, nothing in our universe would be possible. Both are incredible scientific inventions.

Follow a beam or ray of light as it leaves a star or sun and travels through space. This light travels in space at the speed of light in a straight line for up to billions of years unless influenced by a contacted force. Light is a 'universal carrier' loaded full of supplies necessary to create new matter.

However, light is not a Merlin Magician as it cannot create matter all on its own - it requires an engine to spin it. The powerhouse to capture and engineer light into matter is the black hole. Black Holes are our cosmos engines of spin.

As light travels near a black hole, here are the particular forces at work: The black hole contains a spinning helix axis with fantastic revolution speed. As light passes closely by, it is attracted and pulled in. This captured light is then curved so much that the head of the light ray turns sideways and spins like a barber's pole thus creating a spinning helix which pulls and spins the rest of the tail of the ray in behind itself. The result is a spun light ray now in motion as an atom spinning below light's normal speed.

All the properties that light possesses are now available from its tool box: spinning motion; heat; weight; length of strands; thickness of stands; elasticity of strands; a head for the nucleus of an atom; a tail for electrons.

All follow a perfected design pattern for the type of atom element which is being constructed. There may not be an exact blueprint every time. This is a tug a war between the light and the spinning black hole(s). Light wants to stay free to travel, but it is overwhelmed by the forces of the black holes. During the struggle light is pulled, elongated, twisted, turned, and even snapped into lengths. At that instant, the exact type of atom of a certain element is determined.

The weight which light possesses provides 99.99% of the weight and the mass which is found in all matter on moons, planets, stars. Whereas, empty black holes have little weight.

The many special properties of light allow it to be the ultimate recycling substance, going on to infinity, from light to matter, matter back to light. Our universe will continue to grow and expand, being self-perpetuating and designed to reproduce forever. "Nobody has to run out to turn the crank each time a different part of our universe has to recycle."

Thus, matter is made from one material, namely light. The properties of light are many and unusual:

▶ Light is totally indestructible, no difference how many times it recycles.

▶ Light can be sucked in and spun into matter by a powerful black hole.

▶ Light can spin sideways when in contact with a black hole's helix.

▶ Light can form the nucleus/electrons of an atom.

▶ Light spun into matter possesses the propensity to continue to spin as an atom for eons of time.

▶ Light possesses weight providing weight for matter.

▶ Light trapped as an atom of matter always wants to someday escape back to straight line form.

▶ Light will always ultimately escape.

▶ Light carries the property of time within itself.

Light spun into an atom spins as an atom for eons.

Atoms spin for eons because each atom is made from spun light. Light's motto is that it must be ever in motion. (Light is only motionless in its initial storage as the frozen static ocean of light).

Now, one might think the light might come to a grinding halt when a black hole spins it into matter, but not so. Light trapped only forfeits some of its linear light speed as it continues to spin round and round for eons inside the atom.

Also, let's look at the make-up of light for clues. We think of light as this thin long filament whisking across space. But, a ray or beam of light does have a head and tail. The ray can travel forever through space, unless interrupted by gravity.

The light moves in pulses, like waves, as the elasticity of light pulls back to recoil and then pushes forward to keep ahead again. The wave is like a linear pump allowing the ray of light to pulse forward as it travels. When captured inside an atom, light will continue to use its elasticity by rotating.

In the atom, this central axis helix rod within the nucleus is screw shaped and has an elastic property causing it to pump up and down, alternately shorter and longer, which allows it to transfer this up-down motion into a fast spinning motion causing the fat body of the atom (electrons) to spin around the helix. The atom's built-in elasticity allows the pumping motion to continue unabated for eons.

In this regard, the universe is very similar right from the tiny atom with electrons spinning around a nucleus, to a black hole at the core of a sun which spins planets around itself, to the monster-sized black hole at the center of a spiral galaxy which holds millions of star systems swirling around itself. All utilize a very similar *pump toy top* type of pumping action to build continual rotation.

A closer look at the power engine of black holes.

Black Holes are so extraordinary. Imagine the force of attraction required to spin gigantic amounts of light into septillions of tightly packed atoms of elements ever so tightly packed that trapped light cannot escape the tiny spinning confines within the atom for billions of years.

Black holes attain their power from their speed of spin and from the gravitational force that spin creates in a zone around the hole which extends out past its equator. After black holes spin light into matter around their body, they also have that extra added weight and mass to make their gravitational pull even stronger.

Black holes are indeed without equal. They are the most powerful force in our universe. So, what is it about a black hole which allows such power?

- A ferocious spinning mechanism.
- A pump that keeps the spin going.

Black Hole

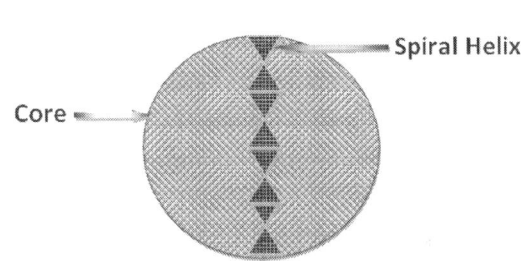

Component parts of a black hole which have to do with spin: interior spiral helix axis is the engine of spin.

The naked black hole axis utilizes an up-down pumping action to create its tremendous bodily spin of thousands of revolutions per nanosecond. This creates a gravity zone.

As nearby light approaches, it is lured. Once entrapped, light displays its fantastic elasticity, as it is stretched out and spun into a tiny tight ball called matter.

The head of the light ray turns sideways and spins like a barber's pole thus creating a spinning helix which pulls and spins the rest of the tail of the light ray in behind itself. The trapped light ray is still in constant motion as an atom but now spinning below light's normal speed. Matter is held so tight as atoms in the black hole's compartments. This ball is tightly spun, occupying only about a 1,000,000,000,000th of the area previously occupied by all the straight line light.

The black hole continues to spin more and more light as it grows in size. Upon this spinning, all the properties of the consumed strands of light are now within the black hole ball. While space surrounding the ball becomes a vast empty weightless byproduct leftover when light vacates the area.

A look at what a black hole can build:

- A **naked black hole** is in the absolute beginning process of capturing and spinning light into its bowels.

- A much **somewhat fuller black hole** has spun more light into matter to form a ball of matter around itself.

- A **full extra small black hole** appears as a tightly-packed moon (extra-small black hole at the moon's core).

- A **full small black hole** appears as a tightly-packed small planet (small black hole at the planet's core).

- A **full large black hole** appears as a gaseous less tightly packed large planet (large black hole at large planet core).

- A **full extra-large black hole** in a solar system will form the largest planet, holding smaller planets and moons in orbit. It will have the loosest atoms of any of the spheres. As this extra-large black hole ages, weakens and tires, its loose surface changes from a gaseous state to a fire plasma state as it becomes that solar system's sun.

- A **full supermassive black hole** at the central hub of a galaxy will outlive all the aforementioned lesser sized black holes which die and then rebuild after the Supernovae which occur within solar systems. These supermassive black holes play by a higher level of rules and Universe Laws than do all lesser size black holes, although many similarities exist.

CHAPTER 8
BLACK HOLE
TO PLANET
TO STAR

Theoretically, every black hole which spins a sphere of matter around itself should be destined to live its long 15 billion year life cycle from black hole to planet (or moon) and complete the cycle as a sun. But in reality, very few of the created spheres are lucky enough to get that much time.

Some suns see their lifespan cut shorter when they are attacked by a rogue black hole. And, the hard-body moons and planets of a solar system can face early curtailment when the sun of their solar system goes Supernova and obliterates them. They speed from hard-body, melting from the heat of the Supernova, and they too explode.

Nonetheless, this chapter follows the successful lifespan cycle of a sun from its start as a naked empty black hole, evolving to a hard-body large planet, and finally evolving to the central orb of a solar system as a sun, ending finally with its death 13-15 billion years or so into the cycle.

Here is a review the major phases in the cycle of the sphere body created around a black hole: short duration beginning birth (CREATION); long midlife with locked up atoms (LOCKUP); a long slow unraveling as we see with our light emitting sun (ESCAPE), a final explosion (SUPERNOVA) of that sun and obliteration of its solar system; and a rebirth (REPRODUCTION). The end of the one 15 billion year cycle quickly ushers in the beginning of the next cycle with double the number of black holes.

Durations of these various phases shown as follows:
CREATION>LOCKUP >
ESCAPE > > >**SUPERNOVA**> >**REPRODUCTION**>
Following is the 17 stages of one 15 billion year cycle taking a naked black hole through to its culmination as a solar system's sun. These 17 stages bring further detail to the 5 main phases presented in the previous chapter.

Stage 1: Naked black hole devouring light and spinning it into matter into its inner bowels.

Stage 2: Black hole spins light to form an exterior body as a small moon/planet which adds more matter, takes shape, adding surface features.

Stage 3: Fire in the central core of the black hole, moon or planet when its oldest atoms begin to unravel.

Stage 4: The interior fire pressurizes, then swells the planet to expand outwards; continental drift.

Stage 5: Earthquakes occur as the lava seeking an escape route pressures the surface: crust buckles.

Stage 6: Mighty inaugural volcano Pimple Burst occurs.

Stage 7: An Ice Age hits the planet.

Stage 8: Planet re-warms after the Ice Age.

Stage 9: The mature black hole at core of planet utilizes its gravity to hold things on its surface.

Stage 10: More atoms loosen, move upwards; surface thins.

Stage 11: Volcanoes increase in number as does the free flow of lava from below to the planet's more liquid surface.

Stage 12: The planet emerges into a small star.

Stage 13: Small star expands to a larger star.

Stage 14: The aging star expands extraordinarily.

Stage 15: Supernova. Star dies as it explodes.

Stage 16: Obliteration. Sun destroys solar system.

Stage 17: Rebirth (Reproduction) (Replication).

Two new back holes are formed for every old black hole.

Each stage of a 15 billion year cycle is further explained: Stage 1: (short stage: elapsed time from hundreds, to thousands, possibly up to one million years).

A naked black hole begins its task with the spinning of matter into its bowels; then, spinning more light into matter to form a ball of matter around itself as a moon or planet.

A naked black hole spins its central axis to attract and spin light into matter. It has incredible spin speed and extremely strong gravitational pull on light. This light spun into matter adds to the black hole's axis length and girth until it is full as a ball of matter. This stage adds weight to the black hole as light carries along with it the property of weight when spun.

The duration of this spinning and building process is dependent on: the size of the black hole; the available easy supply of light; the size and power of the nearby competing naked black holes. Eventually, the ball increases to the size of a small moon or planet. Space, a byproduct, surrounds the sphere as the void which has been emptied of light.

This light which the black hole attracts can be from the frozen static ocean of light, or light traveling from supernova and obliterations, or light traveling from stars. At the origin of our universe, in the first 15 billion year cycle, the black hole(s) were close to the static ocean of light, so a fill-up came easily. Whereas now, a naked hole might find itself far removed from the static light ocean.

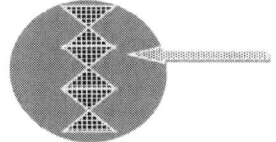

Light is first sucked into the axis and then into the core of the black hole. Once full, more matter is added around the core to form the hard body of the moon or planet sphere.

The sphere now sits in the space it has created around itself with the absence of the light used to build the sphere.

Stage 2: (at 1 million to 2.5 billion years).

Small moon-planet grows larger, takes shape, adds surface.

At this point, the black hole at the core may be 100's of times smaller than the orb it has created around itself. Its core is packed and its surface becomes hard. The surface may display rough features caused by irregularities such as wind, planet rotation, extreme temperatures, and also pits and craters caused by meteoric bombardment from rocks cast into space from other spheres.

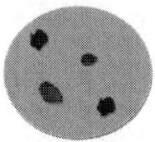

Centrifugal Force of spin causes the sphere to be slightly fatter at the equator. Craters begin to mar the surface struck by meteors fired into space by huge volcanoes know as Pimple Bursts from other spheres.

Stage 3: (at 3 billion years).

Fire in core of black hole begins as atoms unravel.

Every tentacle compartment within the black hole at the core, which spun light into matter, switches from capturing to controlling, from Guzzling to Holding Gravity. As the interior core of a moon or planet, the black hole now spends its total efforts dedicated to control. Eventually, over eons of time, this control loosens in the core, where the oldest spun matter exists. As solid atoms of matter at the core unravel, a tiny fire is slowly ignited deep in the inner core. This matter unravels forming hot molten lava.

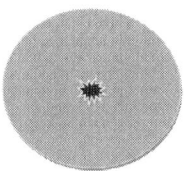

As the sphere becomes right full and increases in size, no more light can be spun into matter. The black hole at the center loses control of its oldest spun light, as matter unravels into a fire.

Stage 4: (at 3.5 billion years).
Interior Fire grows causing further planet expansion.

The interior core fire grows larger when even more atoms unravel deep in the black hole's core. This fire continues to expand moving outwards from the core. Loosened atoms take due to heat expansion take up more area than previously tight densely spun atoms. This ever expanding fire eventually creates so much pressure that the moon or planet has one of two choices: explode, or otherwise find a way to relieve the pressure. Fissures occur up to the surface from the internal pressure allowing lava from the interior to find escape channels heading towards the surface.

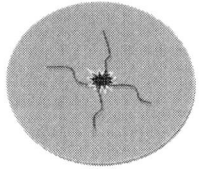

As interior fire grows, pressure mounts causing fissures which the fire can follow to try to get outwards towards to the surface.

Stage 5: (at 3.75 billion years).
Earthquakes occur as the lava pressures the surface.

Trapped lava exerts more pressure causing fissures to become wider and longer to the surface. Internal pressure intensifies as lava crawls up fissures searching for an escape route in order to relieve the pressure. The planet's structure undergoes duress resulting in a large buckling of the surface as gigantic earthquakes occur as parts of matter thrust upwards to form mountain ranges. Also, continental drift occurs on the planet's surface.

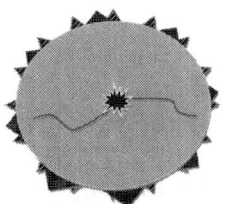

As the lava fissures pressure the surface, gigantic earthquakes cause upheaval as mountains of rock are forced outwards.

Stage 6: (at 3.8 billion years).
The great volcano named Pimple Burst.

Lava, under extreme pressure, becomes trapped in many of the fissures leading to the surface. The entire planet feels ready to explode and destroy itself unless the lava can crack through the surface. At the weakest point, the lava fire attacks upwards through a fissure that is wide. The surface of the sphere buckles upwards from the force of the attack. At this point in the planet's history, one of two things can occur. The planet can explode from the internal pressure, or a large amount of aggressive lava can create a wide fissure to monumentally break through the surface of the planet.

Finally, the initial power-packed volcano named **Pimple Burst** explodes through the surface, packed with so much pent-up lava that the surface juts upwards and the thrust catapults rocks. This first fracture of the sphere's surface is so immensely powerful that rocks are propelled from the surface out past its atmosphere hundreds and thousands of miles into space. They become meteors which can smash into nearby moons/planets to crater the receiver's surface.

Back on the planet, there will never again be a volcano as powerful as Pimple Burst. However, this will be the Age of Volcanoes as many break the surface spewing their lava. New land formations will occur from the lava's pressure. Tons of rock and ash are forced into the atmosphere.

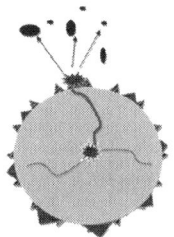

Pimple Burst, the first enormous volcano on the planet swells the entire sphere almost causing it to explode. Pimple Burst is so powerful that it fires rocks like a rocket, overcoming the gravity of the planet sending debris into space.

Stage 7: (at 3.9 billion years).
Ice Age hits the planet.

There may be decades or centuries of cold temperatures right after Pimple Burst as that initial release and Age of Volcanoes spews tons of rock and dust into the atmosphere. While the pressure from the fire at the core has found relief, the surface must face a new challenge as light to the surface is blocked by thick clouds of volcanic ash. This precipitates a prolonged drop in temperature causing the surface of the planet to become frigidly cold. An Ice Age occurs.

After Pimple Burst, the entire sphere experiences a cold Ice Age. An ash cloud blocks out the entry of light.

Stage 8: (at 4 billion years).

The planet re-warms after the Ice Age.

As the cloud cover of ash dissipates, the planet's surface re-warms as light is accepted. Volcanoes become less active; internal pressure from lava pushing upwards on the planet's crust is relieved by lava flowing direct to the surface. The size of the planet has increased like an inflated balloon. As the planet further expands, surface separation occurs as main land masses split apart into continental drift.

As the mushrooming planet expands, Continental Drift occurs.

Stage 9: (at 5 billion years).

Mature planet utilizes gravity to hold surface things.

Determination for the mature planet depends upon its distance from its sun. If it is lucky enough to reside within the Goldilocks Zone, not too close nor too far from a sun, a hospitable primordial soup situation may present itself for life to be seeded. Living organisms on the planet's surface experience the strong holding gravity exerted by the black hole at the core. All upward efforts result in a drop back.

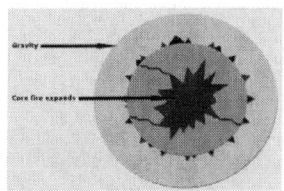

Black hole at the core of planet utilizes gravitational pull to hold matter, even as fire in its core continues expand to the surface.

Stage 10: (at 7 billion years).

The planet's surface thins.

As the central core of the planet loses more of its hold on old matter, and the hot interior fire expands, the interior of the planet's crust is continually depleted and thinned. The surface becomes warmer, then hotter.

The crust of the planet continually thins more from the interior. The overall thickness of the crust is diminished. Anything and everything on the surface feels the increased heat as the internal fire eats away at the tinning crust. The planet sees speeded expansion in size as the interior fire gains in furor. Continental Drift undergoes varying episodes of speeded expansion as continents pull further apart.

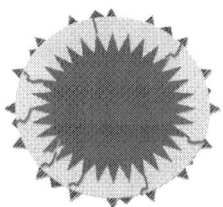

Planet's surface thins as core fire expands reaching the surface.

Stage 11: (at 8 billion years).

Planet's surface becomes more liquid.

The crust of the planet becomes extremely frail and thin as the lava below eats away at it. Volcanoes no longer need to erupt since the surface of the planet had become so thin lava can easily find an avenue to freedom as it flows freely over a large percentage of the entire surface. Temperatures increase dramatically. There is no escape from the heat of the growing lava fields. The melting surface changes from hard rock composition to a soupy liquid. The ground feels like an immense furnace of hot coals. Land masses become unnoticeable. Newly spouted lava no longer cools to add to the hard land surface as in the past. Clouds disappear since a water supply simply isn't available to create a vapor cycle between land and sky. Lakes, rivers, and oceans disappear. Eventually, the planet becomes a fire ball.

This liquid stage for planets provides evidence that the universe works differently than old Big Bang theory had proposed. In sharp contrast, Trillion Theory states that the planet heats up rather than continuing to cool down in the cold depths of space. The fire starts, percolates and grows from the loosening at the core of the black hole. That fire grows outward over billions of years; planet turns fireball.

The planet's surface experiences more of a liquid state as lava from the internal fire encroaches on areas on the surface.

Stage 12: (at 9 billion years).
Planet emerges into a small star.

The crust of the planet disappears as the internal fire melts the entire surface turning it into a fiery state. No longer does lava harden as rock; it now remains hot and molten. The ball has been transformed from a planet, absorbing light and heat, to a small star emitting light.

Planet, with surface fire, emerges into a fiery young star.

Stage 13: (at 10 billion years).

The small star expands to a larger star.

As the fiery star changes all its molten rock and lava from hard matter to a liquid-gaseous state, the star grows in size from a junior to medium sized. Plummets of fiery gaseous eruptions occur from the surface where atoms unravel, escaping to freedom back to straight line light. At this stage, the sun is 4,500 times larger than the black hole at its core.

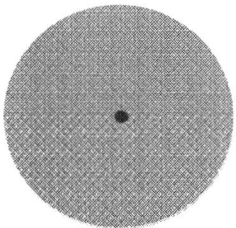

As the black hole at the core of the sun loses control of more atoms, the star expands growing ever larger and hotter. It is now large enough to have a gravitational force strong enough to have a solar system of planets and moons orbiting it.

Stage 14: (at 13 billion years).

The star expands extraordinarily.

With time, the star grows immensely in size maturing from a medium to a huge parent star. Age shows as the star displays erratic behavior and rapid uncontrolled expansion. Bountiful amounts of light freely escape into space. Good fortune might allow the star a long and stately life. However, the massive star is now at its most vulnerable to a perilous attack from any nearby naked black hole.

More light break free after eons in LOCKUP. For a few billion years the sun becomes the solar system's heating house.

Stage 15: (at 15 billion years).

Supernova - the star dies as it explodes.

This elder star grows into a colossal mammoth old star. The oldest stars seem to succumb to old age once they reach 12-13 billion years or so. At the surface of the sun, photosphere helium atoms break apart as light is emitted - freed after billions of years of captivity. Finally, the star loses entire control and explodes providing the universe with one of its brightest sights, Supernova. If there is a rogue naked black hole in the area, the end for the sun may even come sooner from an attack. The black hole's pull on light will speed the unraveling of the sun, creating a bright Quasar, with a rush of light sucked from sun back down into the bowels of the predator black hole.

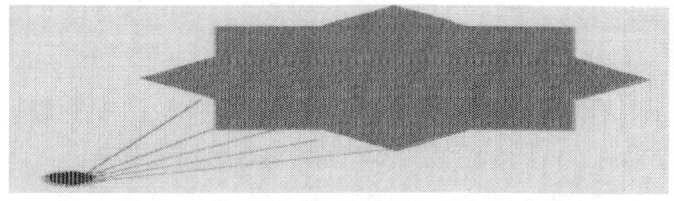

Supernova - the star dies as it explodes.
The surviving black hole can create a brilliant quasar, as a massive rush of light goes back into the black hole.

Stage 16: (at 15 billion years).

Obliteration-The star destroys its solar system.

The sun's enormity soon engulfs the closest planet. Then, Supernova does the rest. Inevitably, the humongous star loses total control going Supernova by exploding all its remaining contents at one time.

Planets and moons fracture and face annihilation as the star destroys most of its own solar system. These moons and planets never get to experience slow growth into a star, for at the point of Obliteration their entire matter is exploded like mega-zillions of nuclear blasts. Planets far away from their sun might survive or be thrust further out into space.

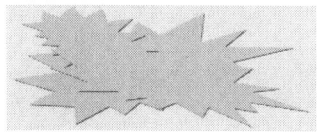

The Supernova power burst from the dying star can destroy nearly every moon and planet in its solar system. This destruction of the solar system is known as Obliteration. Only those further moons or planets furthest away might be spared.

Stage 17: (at 0 years, end of last cycle and start of next).

Reproduction: Two new back holes are formed.

The neutron black hole which was at the core of the exploded sun, plus all the black holes at the cores of the annihilated planets and moons, each split into two new twin binary black holes. In the next cycle, twice the number of black holes will engage in a raging battle to devour light.

Therein, the cycle is complete in one particular zone of the universe, as that zone has renewed and recycled itself yet again. The universe is so vast that many zones unknown to one another may recycle at the same or different time periods. Approximately every fifteen billion years, all zones of the RECYCLIUN Universe do their recycle.

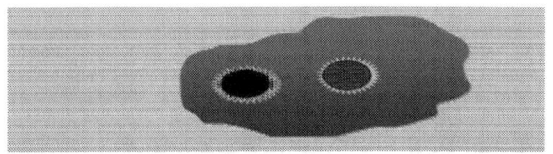

After the Supernova, two new naked black holes split from the original one black hole at the center of the sun, and each planet and moon which is obliterated. This doubles the black hole population with each new 15 billion year cycle

Summary of phases:

Obviously, all of the spheres in our cosmos aren't at the same phase in their life cycle. Each individualistic sphere has its own time frame depending upon many factors. Thus, an entire huge galaxy sees the recycling phases of its solar systems and spheres, at differing paces - not all at one time.

Countless stars and their solar systems recycle on their own cue at 12-15 billion years. Most often these events occur far from other stars and solar systems, so they don't affect another system, although the debris from an exploded planet or a rogue black hole vanquished by the Supernova of a solar system could intrude into an adjacent system.

Imagine this scenario: A calm solar system where the sun is shining and all its planets and moons are slowly accepting light from their sun. All the moons and planets are in their lock-up stage of their existence except for the sun which is in an escape phase. Now enters a rogue black hole, flung there from a different solar system where a Supernova occurred. This entering black hole is now caught by the gravitational hold of this sun, which holds the black hole in an orbit. This black hole is in naked- hungry mode; it attacks the sun of this solar system, madly devouring sun's light, effecting an early Supernova and solar system Obliteration.

This is a strange occurrence, because the sun holds the black hole (its killer) in its orbit, preventing it from leaving, while the black hole devours the sun. The black hole hastens the sun to quickly unravel its atoms of matter back to light.

A long stream of light is seen leaving the sun and then entering the black hole. It is a beautiful sight, this stream of light, so bright and lit-up by the massive amounts of light gushing into the black hole's bowels.

For the black hole, this is a bonanza, a quick fill-up. For the sun, which goes Supernova from the attack, it is a quick death, dramatically lessening its expectant lifetime cycle.

So much light leaves the sun in a rush that the planets also experience over-heating from a sun which quickly mushrooms ever more massive in size. Planets and moons closest to the sun are easily destroyed. Middle distanced spheres are caused to over-heat to their explosion point as their unraveling of atoms speeds up as well. Only the very most distant sphere(s) in that solar system might survive or be flung away across space, and the key word here is might survive. Eventually, the entire solar system, especially those planets and moons closer to the sun, are ablaze, set onto their own path of heading towards Supernova.

As the spheres in that solar system experience their own destruction, the black holes at the core of every sphere are freed of all matter. They become naked black holes again, ready to devour light inside of the remaining graveyard.

Which black hole will get the upper hand by devouring the most light and become the central sphere of a new solar system. This is the survival of the fittest to become the master (Queen Bee) of the new solar system.

It is through the mapping of the life cycles and the many phases of black holes that science can prove Trillion Theory.

CHAPTER 9
BLACK HOLES
BUILT
SOLAR SYSTEMS
(SOLAR SYSTEM FORMATION)

A typical Solar System with sun, planets, and moons.

In our universe's trillion year history, black holes first spun atoms to build individual spheres; then the black holes at the cores of these spheres grouped around suns to form units known as solar systems; and then solar systems grouped around central supermassive black holes to form galaxies. This was the progression followed by black holes during the building; atom world (governed by light spinning inside an atom) > spheres (governed by a black hole at the sphere's core) > solar systems (governed by the extra-large black hole at the core of a sun) > galaxies (governed by the supermassive black hole at the bulging central hub).

In preceding chapters, Trillion Theory has examined how black holes formed atoms and also formed spheres.

This chapter goes back to the early days of our cosmos, to see how the very first solar systems were formed by black holes. This is a most interesting battle and cooperative and between black holes in building a solar system.

Within all solar systems, there exist many small to large spheres. Why this size differential? For example: why are there small planets and large planets, small stars and big stars, why aren't they all just the same size? The answer is that evolution has shown how the fittest grow stronger and larger, and evolution growth is prominent throughout our universe, even in the world of spheres and black holes. For example: The supermassive powerful black hole at the central core of a spiral galaxy evolved that way over hundreds of billions of years as it used its more dominant size to overwhelm a multitude of lesser black holes. It built up an entire galaxy around itself.

So you ask, "Why then isn't our universe just one gigantic solar system or just one super galaxy holding all of the spheres?" Why are there billions of solar systems and galaxies? To answer such questions, we first need to know the origins of the unit solar systems of our cosmos.

So, I put forth a very intriguing preliminary question to make us think of solar system from a strategic point of view; is our solar system a cooperative or a battle scene?

Most people after some thought would normally answer, "A **cooperative,** because the sun, planets and moons in our solar system appear to exist harmoniously, helping each other out in numerous ways and there appears to be no fighting going on between any of the spheres."

I would answer back, "Yes, correct right now, as our solar system is presently in a cooperative phase between all of the spheres at this point in the 15 billion cycle of our solar system. However, that was not the case when the previous solar system of our area died and the battle commenced by all the black holes of this area to rebuild, resulting in our present new solar system."

It is during this **battle** scene between naked black holes, fighting against other black holes with the goal of pulling in energy and converting it into matter around their bodies, which determines the structure of all new solar systems.

The battle scene.

First and foremost, it is a battle between naked black holes in a warlike phase before a solar system is formed. Naked lack holes fight against one another to devour energy into matter. This battle phase occurs at the end of one solar system and continues on into building a new replacement solar system. This battle is a short phase, lasting just a few decades. This gargantuan battle occurs between the newly exposed naked black holes which were at the core of every moon, planet, and sun (naked because their matter returned to energy and escaped their surface during the supernova of the sun of their old solar system). Holes, now unburdened by matter of their surfaces, regain their fast powerful spin as they struggle against nearby opponent black holes for their strongest position in the next new solar system. This battle will rage until the entire structure of the new solar system has been settled and established and all the full black holes enter the subdued cooperative phase of this new 15 billion year (approximate duration) cycle of the new solar system.

The forming structure of this new solar system follows the universe rule of 'size matters' as the largest black holes win the battle for light thereby forming the largest bodies around their sphere and they possess the strongest most far-reaching gravitational pull. The largest strongest black hole (sphere) forms the central sun and the next largest black holes form the planets, while smaller spheres comprise the moons. Note: Universe Laws prevail in setting the new solar system structure. No moon can outsize the planet it orbits since size determines gravity strength and dominance.

The cooperative scene.

After the battle between black holes ends, forming the new solar system, a **cooperative** exists between spheres in a peaceful phase when each sphere in the solar system is full.

Cooperative is the much longer phase than the preceding battle. In the cooperative, the built mature moons, planets, and sun(s) of a solar system co-exist in harmony with one another. The order of rank within the cooperative was determined by the gravitational power size of the black holes at the core of each sphere and proximity distance between spheres plays a secondary role when determining a spheres positioning within the new solar system.

The black hole at the core of each sphere (moon, planet, or sun) provides the powerful spin, which provides a sphere its gravitational pull, and in our present solar system these various gravities peacefully cooperate. Example: our planet holds our cooperative moon in orbit and our entire solar system appears cooperative as 8 planets orbit our sun's gravity and each planet holds moons in orbit depending upon each planet's size and gravity strength. Over 100 moons have been catalogued in our rather large system.

So in review, our universe is as much a fight between black holes as a cooperative between spheres. The battles between black holes, all fighting to attract energy and convert it into matter around themselves, have provided the organizational structure for each and every solar system. All of this battling followed by a cooperative is governed by Universe Laws which operate between naked black holes and thereafter between black holes existing at the cores of the spheres which they built around themselves.

To view the first solitary solar system which ever existed in our universe, we need to travel back one trillion years.

First tiny simple solar system a trillion years ago began in the first 15 billion year cycle.

Trillion Theory shows the Universe Laws which the first solar system obeyed at start up. A trillion years ago, an ocean of light, limitless in strands in every possible known direction, occupied the entire known physical cosmos,. This light ocean was frozen full of static light. At this point, there weren't any such things as spheres or space yet - only the endless ocean of static light. (Recently, scientists brought light to a standstill, demonstrating static light is possible).

An ocean of frozen static (non-moving) light was all that existed at the origin of our universe a trillion years ago.

Right thereafter, the first black hole appeared in the ocean of light, providing the prime historical moment when the action of our universe really began. That initial black hole possessed an axis which spun at tremendous velocity. As the engine of the universe, that first naked black hole utilized its powerful spin to loosen frozen static stands from the ocean. Once freed, those light stands overcame inertia, and then quickly accelerated up to the speed of light. But, before they could make their escape, that's when the light rays were most susceptible to an immediate pull from the naked black hole. Captured, the light strands were spun into atoms of matter deep into the bowels of the black hole. More light was gobbled up as the black hole formed a body of matter.

The first constructing CREATION phase had started with the building of the universe's first sphere. Eventually, the black hole completed its spinning of light and was now as a small planet in LOCKUP phase - holding matter in place.

Space had been created around the orb as an empty byproduct left over from the disappearance of millions of tons of light from the static ocean of light. In simple terms, the black hole had eaten a cavern of space inside of the non-moving ocean of light. The devoured light, spinning tightly as matter around the sphere, now occupied less than a trillionth the area of the static light it had devoured.

The first black hole was introduced into the static ocean of light. It loosened static light from the ocean, freeing that light and spinning it into matter around itself, leaving empty cold dark space in the area close around itself. As that first black hole continued to devour light, it grew into universe's first planet.

Next, a second black hole was introduced into the static ocean of light, near the first black hole. The second younger black hole revolved around the first black hole which had grown to a planet. That planet's strong gravitational efforts to hold its contents in place extended outward to keep the young black hole in orbit. The second black hole attracted and spun light into matter from the static ocean of light, but it also hastened the loosening and expanding of the planet.

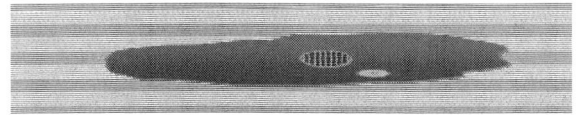

A second black hole was introduced, orbiting around the already formed planet (larger sphere). First tiny solar system formed.

The small naked black hole ate away at more available light from the ocean of light, but its ultra-fast spin also pre-loosened the atoms on the planet. The planet's loosening atoms hastened its transition from a planet with a hard surface to an expanded, hotter planet with a more liquid surface. The unraveling planet, which now had atoms loosening as it was losing control, saw a steady unraveling of its atoms creating a stream of light leaving its own surface and being devoured by the new black hole. The first sun, with a fiery surface, was the result.

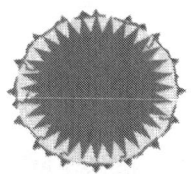

The planet's transformation to the star stage was hastened.

Eventually, the sun loosened (expanded immensely) while it still held the eating black hole in orbit. Therein, the first solar system was born with one central parent sun (with the original black hole inside its core) and one small orbiting childlike planet (with the other black hole inside its core).

Over the coming eons, the new small planet devoured more light from the sun further loosening and expanding the aging sun. Ultimately, the sun over-expanded and went Supernova and Obliterated the planet which it held in orbit. As the process completed, the sun rid itself of its matter which unspun back to light. The orbiting planet was burned to a crisp (obliterated) by the exploding sun such that all of the planet's atoms of matter exploded and went back to light. The first solar system came to an abrupt ending.

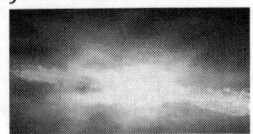

First the sun went Supernova and exploded.
The Supernova effect melted the planet into Obliteration.

This may seem strange for the star to extend gravity out to the young black hole to keep in orbit. All the while, the young black hole freely preyed, devoured and destroyed the star. But in the end that, turnabout was fair play. The star said, "Yes, you can eat me, but I will hold you in orbit, and when I increase tremendously in size and I'm forced to explode, your matter will be annihilated too."

On the heels of the Supernova and Obliteration, there was no evidence left around as proof showing that the first solar system had ever existed. No long-term graveyard or markers recording its history. The recycle left no historical fossils.

Except that is, for the two surviving black holes: one had been at the core of the sun; the other had been at the core of the planet. The sun's black hole had survived going Supernova and the planet's black hole survived Obliteration. Universe Law states a black hole can never be destroyed.

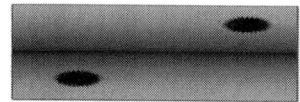

**Two naked black holes survived Supernova
and Obliteration of our universe's first solar system.**

The second solar system (2nd 15 billion year cycle).

Black holes can never be destroyed. Thus, the two black holes of the first solar system survived a Supernova.

These two inaugural black holes each then subdivided via thereby increasing their numbers from two to four.

The four new naked black holes of the region immediately commenced a battle for light from the nearby static ocean of light and also from the light which was attempting escape from the sun's Supernova and the planet's Obliteration. Each naked black hole spun on its central axis, spinning light into matter. Each black hole devoured light adding to the length and girth of its central axis. They then added to the size of their cores. Each of the four new black holes in this second cycle was larger than the two black holes of the first cycle.

Of the four new naked black holes, the one being the best positioned and most opportune consumed the most light and was the first to shift its main focus from consumption of light to the long task of holding matter. It also gained the most overall mass of the group holding the other 3 in orbit. It was destined to grow into the sun of that new system.

In the long run, this second solar system faced the same fate which had befallen the first solar system. In solar system two, the three spheres held in orbit by the largest sphere were still in devouring mode. They loosened the atoms of the central sphere which was already in control mode. After billions of years, that central sphere loosened from a large planet to star. After more billions of years it loosened as a star and went Supernova. Likewise, that second solar system also met with Obliteration. The devourers of the central sun caused it to go Supernova and it played the same dirty trick on its devourers as all were obliterated by the Supernova.

Of course, the Universe Law once again prevailed as black holes at the core of the sun and planets survived. There were four surviving black holes after Obliteration of the second solar system. Each was already in motion working hard to reproduce by fission and split into two new naked black holes, upping the count of naked black holes from 4 to 8.

The third Solar System (3rd 15 billion year cycle).

In the third solar system, cycle three, there were 8 new naked black holes. The sphere population had doubled again, ready to form yet again a new larger solar system with one central sphere and seven spheres in orbit. With 8 spheres in play, the Universe Laws which existed between black holes, and planets, and suns, were put to a sterner test.

As usual, the black hole which was the largest dominant eater of light became the focus of the next solar system. Over the next few billions of years it grew enormously from being the largest planet in that new system to being the sun.

Other changes occurred as well: The ocean of light was now further from the grasp of any of the black holes which were growing their bodies into planets. Thus, the new solar system had a more spread out appearance. While the central largest planet (sun) held the other seven black holes in orbit, the black holes were pulled outwards by the attraction of the light ocean on the perimeter. Expansion of the universe was in full swing. More light was consumed by each larger solar system which stretched itself further from the center of the cosmos to stay close to the perimeter's ocean of light.

A secondary dominant player appeared in system 3. While the largest planet became the sun of the new larger solar system, that sun had difficulty controlling each and every one of the planets in orbit in its solar system. One of the larger planets exerted its powerful gravitational pull on a smaller sphere forcing it into orbit as a moon. So, this 3rd solar system had the structure: 1 central sun; 6 planets; 1 moon = 8 spheres. At the end of 15 billion year cycle three, the Supernova and Obliteration were more explosively super because this was a larger sun than the past suns.

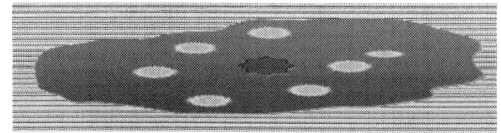

Solar System 3 (8 spheres). The static unmoving ocean of light on the perimeter surrounds empty space which the spheres had created by consolidating light into matter around themselves. One central sun, six orbiting planets, plus one small planetary moon, comprise the 8 spheres in solar system 3.

The fourth 15 billion year cycle.

In cycle four, the count of black holes doubled again, this time from 8 to 16. This time the black holes had been blown even further apart from one another by Supernova's extra power. There were various possible formations.

Scenario one: one new larger system comprised of 16 spheres. With that many spheres different formations were likely: one center sun and 15 orbital planets; or one central sun holding planets in orbit but some large further away planets holding moons in orbit around themselves.

Scenario two: The two largest survivors from the old sun going Supernova are the two contenders vying to be 'Queen Bee' as the central force(s) of the new solar system. Two spheres control the solar system becoming its binary suns.

Scenario three: The two largest surviving twin black holes from the Supernova are driven far apart by the explosion's power. The less dominant black holes followed either of the two leaders, forming two new solar systems.

Scenario four: An unequal number of black holes were carried away with each of the two dominant black holes which were flung apart. One dominant black hole ended with 12 spheres while the other dominant black hole ended up with only 4 spheres. In this cycle there are two new solar systems; one large and one small. The larger solar system, with more mass, extended a greater gravitational pull and held the smaller solar system in a distant orbit. As a direct consequence, the universe's first small galaxy formed.

Some Solar System Rules (Cosmic Universe Rules):

• Holding gravity of a sun to keep planets/moons in orbit depends directly upon its mass. Gravity G, the gravitational constant of the universe, times the mass of the orb, divided by the square of its radius. Gravity = $G (M/r2)$.

• Pimple Burst always occurs on planets/moons. As a sphere

begins a transition from devouring light to holding matter in place, there eventually comes a time when the onset of loss of control of matter will begin at the core of the sphere. This will initially cause lava in the core of a planet/moon to search for an exit. As eons pass, this pressure intensifies such that the sphere can either explode or find a way to ease. Usually lava finds a fissure out to the surface where the initial Pimple Burst eruption will have such titanic force that the fracture propels massive amounts of debris into space. to pelt down and crater onto other spheres.

• The distance between spheres becomes greater with each recycle into a new solar system.

• Black holes of new solar systems always grow larger than preceding black holes of past systems.

Each of the subsequent 15 billion year cycles doubled the universe size and population.

As each of the 15 billion year cycles of the universe took place, each brand new cycle saw the sphere population doubled and space widened within each new solar system or in larger venues called galaxies.

But, not all of a galaxy recycles at any one time. Galaxies are so enormous that their solar systems are totally remote from one another. Approximately every 15 billion years, all zones of the RECYCLIUN universe recycle, thereby doubling the sphere population in that zone, thereby contributing an ever increase to the size and population of our cosmos.

****

OUR SOLAR SYSTEM FORMATION

In our current solar system, we are in a long peaceful **cooperative**. A few billion years from now, when our sun goes Supernova, explodes and destroys our solar system, the **battle** of all the naked surviving and newly reproduced black holes will commence with the goal for every black hole to rebuild a new sphere around itself. This battle will see a whole new restructuring of the next new solar system. The battle may even see a powerful split with two new solar systems replacing our old one. As a result, our Milky Way Galaxy will gain more solar systems and spheres. Our universe is alive and growing. Trillion Theory shows that our solar system, with our Earth as the third rock from the sun, presently resides in the 67th of the 15 billion year recycles of our cosmos, with 73 quintillion stars in that cosmos.

In our universe's trillionth birthday, our solar system is only one of billions making up our mammoth Milky Way Galaxy. And, Milky Way is just one of billions of galaxies.

Now, we revisit the start of our present solar system, 10 billion years ago. The previous 66th cycle had just ended.

So, when our new solar system was replacing the old solar system which for 15 billion years in cycle 66 used to occupy this area of our Milky Way Galaxy, a percentage of other solar systems within our galaxy were doing likewise. Therein, the overall population of Milky Way was growing.

When the old previous solar system in this locale aged out, the sun at its center went Supernova. Universe Laws determined how that Supernova occurred, the effects it had, and what happened right after, thereby controlling how our new solar system formed and replaced the old one.

When that old aged star went Supernova, the whole old solar system was affected. The old sun had overexpanded to initially engulf and melt the planets closest to it. Finally upon Supernova, the old solar system faced Obliteration.

Chain reaction events occurred for the old solar system:
Step 1: the old sun shed all of its matter at one Supernova moment, releasing it back to straight line light.
Step 2: the black hole at the core of the sun spun hard in an opposite reaction and split into two new naked black holes.
Step 3: these two new black holes were driven far away from one another, receding away from the force.
Step 4: the two black holes began their battle to devour light escaping from the area. This is the glow pouring back into the black holes and appears as an implosion of the old star, but in reality is just the new naked black holes pulling the light which is trying to escape, back into their bowels.
Step 5: all the planets and moons in the solar system were destroyed shedding their matter back to light as they also went Supernova thereby splitting their black holes cores.
Step 6: half or so of the smaller black holes followed one or the other of the two major black holes created from the Supernova of the old sun. Two new solar systems resulted and accelerated away from one another.
Step 7: a new solar system began to form from the black holes now occupying the area of the old solar system.

Steps occurred in forming our present new solar system in the 67th 15 billion year cycle of our physical universe:
    Now that the old solar system at the end of cycle 66 is gone, we follow how our solar system formed at the start of cycle 67. Universe Law will guide us by showing why we have a central sun; why planets and moons are located where they are; why our solar system is quite linear in shape; why all the planets rotate in the same direction around our sun; why the different spheres have different sizes, axial tilts, and surfaces; and why planets rotate on their axis. Basically how our whole solar system developed to the present.

Our Sun, an x-large black hole formed into our Sun.

The x-large black hole which would become our sun (which was also at the center of the old sun of this area) was the first naked black hole ready after the Supernova to begin devouring light. The other black holes at the center of the old planets/moons were just beginning to lose their atom contents and when they did, the large black hole was ready to devour the emitted light from their unraveling bodies.

That x-large black hole was the most gluttonous greedy voracious eater of light. It became the Henry VIII of our solar system. Basically eating until it was ready to bust. It hogged light, using its size, early positioning, and gravitational pull to steal light from the grasp of others.

As that x-large black hole captured light shed from the other unraveling spheres of the old solar system, it stored the new matter deep into its bowels. This light entered the x-large black holes bowels as 'coming in hot' (light carries the heat property). So much light was being consumed by the x-large naked black hole in a short period of time (Quasar) that most of the atoms were 'staying hot.' The earliest 'hot' spun atoms were covered and layered over and kept hot by the next blanket of atoms spun over them.

Our sun is much larger than any of the planets because: it has the largest black hole at its core; that black hole spun tons of light in short time which it captured from the nearby source of the dissolving planets and moons; that light 'came in hot' in abundance and stayed hot making for the loosest spun atoms (occupying more area) of any of the spheres in our system. (Universe Law: the larger a black hole the looser it spins its atoms, layer by layer, from inside to outside).

As the largest black hole of our solar system, our sun gained the most mass and the central position as it held the other smaller black holes in orbit. But, as that black hole gained mass and it grew into a sun, its mass empowered it to hold all the forming planets in more permanent orbits.

While the x-large sphere utilized its powerful gravity to hold the other spheres in orbit, the other smaller black holes were still in the eating guzzling stage and they attacked the loosened up x-large sphere which was holding them in orbit. Their attack further loosened the atoms of the x-large sphere turning it into a sun to unspin and emit light.

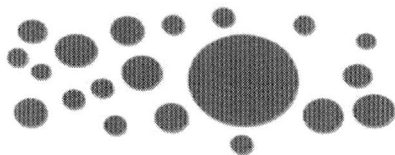

When a star goes Supernova and its Obliteration destroys the old solar system, all that is left behind are naked black holes, poised and ready to rebuild a new solar system.
The graveyard left behind shows little of the past spheres. History is mostly wiped out. But, new naked black holes survive and they begin the search for new available light to spin into matter around themselves. At this stage, the black holes are in a scattered pattern. Axial tilts can vary with each black hole which searches for the best position to attract and devour light. That light could come from the static ocean of light if nearby, or from any light that can be captured before the light from Obliteration escapes, or there can be the entrance of light from distant stars.

Planets in our Solar System.

Planets in our solar system play second fiddle to our sun. Since our sun spun most of the available light in our new forming system into itself, all the lesser naked black holes orbiting it struggled to find light to spin into matter.

These smaller naked black holes entered into battle with our sun, which had become so full that it had moved from spinning light to lock-up while trying to hold its contents.

The naked smaller black holes were spinning fast, looking to capture light, and the gravitational pull of their spin was further loosening atoms on our sun's surface.

The smaller naked black holes pulled much light from the sun, spinning that light around their own cores. They would have forced our sun to go Supernova right then and there except for one thing: as they became fuller forming into planets and moons, their eating slowed to a trickle.

Thus, our solar system was saved and moved to what it is now a sun steadily emitting light and planets and moons accepting that light to warm their surfaces. These lesser spheres are unable to spin anymore of that light into matter.

Our sun spun so much light into matter so quickly and that matter 'came in hot' in such large volumes that our sun forfeited a chance to have this matter cool.

In contrast, our planets and moons spun their matter slowly. Sure, the light 'came in hot' but each new layer had more time to feel the coldness of space before more hot could be spun overtop the previous layer. Therein, only the center of planets/moons has remained hot. However, even for planets/moons that internal hot always grows in size as more atoms loosen on the interior. That's why hot lava from below always fights to the surface. Basically, the loosening of atoms in the interior always eventually wins out over the coldness of space. Thus, a planet is never solid cold, because the hot central core always percolates.

Organization of our Solar System.

The organization of our unique solar system followed a set of quite common Universe Laws. Our solar system has a sun with 8 planets which are listed below from closest to furthest from the sun, with diameters and number of moons:

Mercury (3,032 miles in diameter)(no moons)

Venus (7,521 miles in diameter)(no moons)

Earth (7,926 miles in diameter)(1 moon)

Mars (4,222 miles in diameter)(2 moons)

Jupiter (88,846 miles in diameter)(63 moons)

Saturn (74,898 miles in diameter)(62 moons)

Uranus (31,763 miles in diameter)(27 moons)

Neptune (29,700 miles in diameter)(13 moons)

Total of 177 spheroids; 1 sun, 8 planets, 168 moons. Why are the spheroids where they are? Accidental? No. Universe Laws tell us smaller spheres were more easily pulled close to our sun. Whereas, larger spheres stayed further away.

Organization followed Universe Laws pertaining to black hole sphere size and gravity pull. Mercury (smallest) was the sphere most easily pulled close by the sun's gravity. Mercury was too small to hold orbiting moons. Venus, also no moons as it was overpowered by gravity from the close in proximity sun; Earth 1 moon and Mars 2 moons are a little further away from our sun's powerful gravity. Then, the gas giants were large with enough mass to hold many moons and challenge the distant sun's gravity; Jupiter has 63 moons; Saturn 62; Uranus 27; and Neptune 13. Saturn is the most interesting planet with its rings circling around its equator, illustrating universe gravity laws around a sphere.

Our solar system displays a rather flat pinwheel shape. Our sun is the center pin and the planets orbit at varying distances from the sun. These planets are in a flat plane because of the Cosmic Law stating that black holes, and the spheres formed around them, exert their strongest gravity pull extended out past their equator.

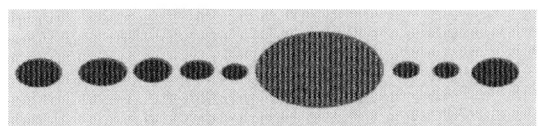

Our formed solar system took on a linear shape because of the black hole at the central core of our sun. The spin of that black hole rotates the sun, but also creates a gravitational pull holding the planets in orbit extending out along the sun's equator.

Universe Laws which spheres in our solar system obey:

After a black hole fills up from the eating light, its gravity powers change from the spinning light to the holding of manufactured atoms in place. This gravitational pull leads outwards from the black hole at the core of the sphere and extends out past the surface of the planet in a linear flat plane extending out around the equator of the black hole sphere. The larger a sphere and the greater its mass, the further its gravity extends to hold more spheres in orbit.

The x-large black hole at our sun's center (largest in our solar system) possessed the strongest gravitational powers to win the battle to attract light, spin it into matter, and build a massive body around itself. Once that task was completed, it began exerting its strongest gravitational force on the other lesser black holes pulling them into orbits around itself. As eons passed, and all the lesser black holes spun more light into matter, planets formed around these lesser black holes orbiting the sun, and moons formed around the still lesser black holes orbiting large planets.

The orbital direction for the 8 planets of our solar system followed the Universe Law stating that all planets held in a sun's gravitational pull revolve around that sun in a direction determined by the direction of spin of that sun. Our sun spins counterclockwise on its axis, thus all the planets orbit it in a counterclockwise direction around our sun.

However, moons are first and foremost tied in orbit to the axial direction of spin of the planet they orbit. That orbit could be clockwise or counterclockwise relative to the planet depending upon the direction of axial spin of their planet.

Note: Here, it is important to differentiate that spin is different than orbiting. A planet rotates (spins or pivots) on its own axis, yet it can revolve (orbit) around our sun. The main motion that a sphere has is the rotation on its own axis. The direction and speed of that rotation is caused by the direction and speed of rotation of the black hole at the sphere's core. Whereas, movement of the sphere in orbit is determined by the gravity cast by the largest nearby sphere. Example: Our moon spins counterclockwise on its axis since the black hole at its core spins counterclockwise. But, our moon revolves counterclockwise around Earth because of the counterclockwise spin of the black hole at Earth's core.

Next, here is where both Big Bang and Nebular Theory are said to be wrong by Trillion Theory. If a circling nebula cloud caused our solar system, then all the planets should have been imparted with a similar direction of rotation on their axis. Because our sun has a counterclockwise axial spin, then every planet should have counterclockwise axial spin.

But this isn't so. All the planets and moons in our solar system don't spin the same direction on their axis. This is so because our solar system didn't form from a swirling nebular cloud. Trillion Theory shows that spin direction for a sphere is totally dependent upon the direction of rotation of the black hole at its core. This direction is determined when a black hole begins its cycle. When a black hole splits during replication, the split sets the two new black holes into their direction of rotation. In our solar system, Mercury, Earth, Mars, Jupiter, Saturn, and Neptune rotate counterclockwise on their axis, just like our sun. But, Venus and Uranus don't.

Weirdo bizarre Venus and Uranus spin CLOCKWISE. They make a nightmarish unexplained mystery for astronomers. They cast substantial doubt on Big Bang and Nebular theory.

That mystery is best explained here in Trillion Theory. Both Venus and Uranus, the so called weirdo planets, simply have black holes in their cores which spin clockwise. (Note: The direction of spin of spiral galaxies has been found to be 50-50, indicating that half the supermassive black holes at the hub of galaxies spin counterclockwise, half clockwise).

Now, Universe Laws within Trillion Theory even deal with the speed of spin of spheres on their axis. Of course, this is tricky since the speed of spin of a sphere on its axis changes over time. It is most likely that black holes, small or large, possess the propensity to spin at the same high speed initially when they are naked black holes spinning light into matter. But, the size they grow into, the contents which they control and the influence of other powerful black holes in the vicinity can affect their rate of spin on their own axis.

In writing 'Rotational Speed Laws,' Trillion Theory keeps them separate for naked black holes and spheres with mass:

Rotational Speed Laws for Black Holes:
• Naked black holes have a super-fast rotating speed.
• Fuller black holes witness slowing of their rotational speed.

Rotational Speed Laws for Spheres:
• After a black hole body builds a body, its spin slows.
• Partially full black hole: sluggish speed.
• Full spheres: even more sluggish speed.
• Suns are more sluggish than planets.
• Larger suns are more sluggish than smaller suns.

Proximity Rotational Speed Laws for Spheres:
• Close proximity to a larger sphere slows rotational speed of the smaller sphere.

Hopefully, this chapter fostered a greater understanding of how our solar system came to be, how it operates, and why some things in our solar system are the way they are.

CHAPTER 10
BLACK HOLES
BUILT GALAXIES

Spiral galaxies are a near perfect organized island colony.
Trillion Theory introduces a new understanding of galaxies.

Galaxies.

Galaxies are definitely the most ominous beautiful sights in our entire cosmos. They present themselves with power, magnitude, majesty, and magnificence as they show off their splendor and flaunt the beauty of their spectacular array. Galaxies are wondrous yet solitary. They can be described as gigantic remote private islands of the cosmos. Our present cosmos is estimated to hold two hundred billion galaxies. One galaxy may contain millions of stars with solar systems.

While solar systems recycle on average every 15 billion years, galaxies are much more permanent features because they don't live and die under the 15 billion year rule. To clarify: it is the solar systems within a galaxy which recycle at various times during every 15 billion years. Galaxies remain more permanent and grows in size each time one of their solar systems recycles; they either gain larger solar systems, or more of them, or both. In any particular 15 billion year cycle, various solar systems in the different quadrants of a galaxy may recycle, fostering galaxy growth.

So, think of a galaxy as one of the many island hotels of the cosmos. A galaxy may have millions of hotel guests residing as solar systems. No matter how many times the guest occupants (solar systems) of the hotel rooms change, that remote galaxy hotel remains intact as the main entity which gains in size and number of rooms each time.

Once a galaxy becomes gigantic, its configuration easily survives a Supernova recycle of one of its solar systems while the rest of the galaxy remains intact and unaffected. While various parts of a galaxy recycle every 15 billion years, the galaxy itself may be hundreds of billions years old.

The oldest largest galaxies have gone through as many as 30 to 50 of the 15 billion year recycles of their contents, over the past 400-800 billion years. Estimate the number of stars in a galaxy and it is possible to calculate the age of that galaxy using the doubling premise each 15 billion years. Each cycle, the galaxy grew larger in stellar population.

Galaxies are the absolute best time recorders of our universe's history. What wonders they must have witnessed over their hundreds of billions of years.

Spirals make up about 90% of our universe's galaxies. Their stars and solar systems are in orbit whirling around a galactic bulging center. This bulge has enormous mass with enough gravitational power to hold in orbit millions of stars and solar systems. At the core of the bulge is a fast spinning supermassive black hole middling a large concentration of densely clustered superstars. That is where Cosmic Laws pertaining to black holes are rewritten.

But, don't think of a galaxy as simply a huge solar system. A solar system has spheres orbiting a sun; whereas, a galaxy has many solar systems orbiting a supermassive black hole.

And, in a galaxy, the supermassive black hole centering the galaxy has evolved to a new level of existence.

A supermassive black hole centers a spiral galaxy.

This chapter will show the evolution of a black hole which becomes the monstrous sized central bulging core of a galaxy. Universe Laws are tweaked by the supermassive.

History building up to galaxies:

A trillion years ago our cosmos started from one small solar system growing into more, and eventually into billions of galaxies. We detail key 15 billion year recycle points:

• The first solar system had 2 spheres (sun and planet)
• That first solar system grew in size entering cycle two
• The second solar system grew in size entering cycle three
• The third solar system grew in size entering cycle four
• The 4th solar system, split into 2 solar systems for cycle 5
• By cycle 6, there were 4 solar systems in total
• The largest sun kept all the other solar systems in orbit
• That formed the first small galaxy of our cosmos
• At the center of that galaxy was a massive sun
• At the core of the sun was a massive black hole
• The massive went Supernova and Obliterated the galaxy
• Two new galaxies formed replacing the old galaxy
• Black holes at cores of galaxies became ever more massive
• Massive black holes evolved to being supermassive
• Super size brought structural changes to the black hole
• These supermassives evolved new laws for themselves
• These supermassives learned how not to go Supernova
• These supermassives learned how to become ageless

Evolution from solar system to galaxy.

Trillion Theory has shown how our universe began small a trillion years in the past. Black holes ate from the static ocean of light forming the first solar system. This one system grew with each recycle into several solar systems. As several solar systems came under the control of the central massive star of the largest solar system, the first galaxy formed. As galaxy recycled, it split apart forming two galaxies. Today, we see billions of galaxies containing solar systems.

During this progression in numbers, the main change which occurred was the size of the most dominant sun of the largest solar system. Every 15 billion years the massive sun centering the largest solar system grew ever larger, capable of holding evermore solar systems in orbit.

However, the true progression which occurred was that it was the black hole at the centre of the massive sun which grew proportionally more dominant each cycle capable of forming a very massive sun. And, that black hole graduated to a supermassive size, capable of controlling a galaxy.

At a definitive point in its evolution, this supermassive black hole determined not to eat light solely for the building of a supermassive sun around itself – that was too mundane and unproductive. If it did so, it found that it was forced as a sun to go Supernova and detrimentally destroy its galaxy. A smarter strategy presented itself, allowing the supermassive to evolve much older than just one 15 billion year cycle and last forever centering its galaxy. That way, the supermassive black hole could grow its galaxy over 100's of billions of years. To complete such a feat, the supermassive black hole needed to change its strategy for conducting its business.

How did this occur? Perhaps a supermassive black hole was preprogrammed to change the rules it played by when achieving certain benchmark size and duties.

Galactic Obliteration.

What is Galactic Obliteration? Why is it relevant?

When the galaxy is still small, the humongous star that centers the galaxy can go Supernova and make every star and solar system within its galaxy face Galactic Obliteration, death and rebirth of the entire small galaxy. This was by far the most brilliant event in our cosmos when a galaxy went ballistic; the sight of a multitude of Supernovae exploding at one time far exceeded the brightness of a single Supernova.

In certain ways a Supernova creating Galactic Obliteration was good. It was a way for an entire small galaxy to instantly double its number of solar systems.

The bad was that after the Supernova by the massive sun centering the old galaxy occurred, too often that powerful Supernova split the galaxy into two, pushing apart the two largest massive black holes which had centered the massive sun. Each of the two massive black holes took half the lesser black holes along to form a new galaxy. Thereby, two new galaxies receding away from each other replaced the old galaxy. Thus, it was difficult for any one galaxy to remain intact and grow ever larger.

Also bad was that after each of gigantic Supernova a battle raged by all the massive black holes during Galactic Reformation to see which one would become the master (Queen Bee) of the new galaxy. With Queen Bee changing every 15 billion years, any newly crowned Queen Bee had to discover a method to remain on the throne as long as possible as a supermassive black hole centering the galaxy.

The good was that sometimes after Galactic Obliteration, the galaxy stayed as one, and grew ever larger. Its remote solar systems went Supernova at separate times without affecting the central dominant massive black hole.

Each new Galactic Reformation saw the central black hole more disproportionally large in ratio compared to all its other diminutive rival black holes. Eventually, the prominent master black hole (Queen Bee) had hundreds, to thousands to millions of solar systems in its galaxy. So strategy-wise, destroying all at a moment's notice was not productive for the Queen Bee supermassive black hole. That wasn't sound strategy, and black holes are all about strategy. And the larger a black hole becomes the more cards it holds and the harder it can trump any other smaller black holes.

Finally, the point came for the supermassive black hole (Queen Bee) where it wanted to maintain that which it had built. The definitive strategy for the supermassive black hole was to adjust its structure giving it new powers. These new powers allowed the Queen Bee supermassive black hole to overcome the rule which took black holes into the recycling process every 15 billion years.

Breaking the rule of 15 Billion Years. The supermassive black hole (Queen Bee) of a large spiral galaxy changed the rules allowing itself to grow far older.

All moons, planets, suns and solar systems recycle every 15 billion years in the Recycliun Universe, and initially so did the massive suns centering galaxies. But, for any sun to avoid going supernova, the massive black hole at its core had to develop new strategies. The cornerstone of this new strategy called for the massive black hole to be able to control the galaxy without having to grow into a massive sun itself, and then later have to go Supernova. Note: It is suns which every 15 billion years weaken and go Supernova. Whereas, the XL black hole at the sun's core always survives.

So, the soundest strategy for a massive black hole at the center of a galaxy was to avoid having to grow into a sun. Step one in this new strategy (breaking Universe Law) was to do something totally different with the tons of light which the massive black hole pulled inwards from other suns. This required a structural change for the massive black hole by evolving structural transformations.

A quick review: Only a small percentage of black holes ever evolve enough to make it to massive status. First, a black hole must become dominant as a sun in its solar system. Then, as a next step become the dominant XXL black hole at the core of a sun in control of a group of solar systems in a small galaxy. With each of these successful steps, the black hole grows in size and dominance. Each step, its central helix becomes longer and the core of its body increases in size and in the number of compartments.

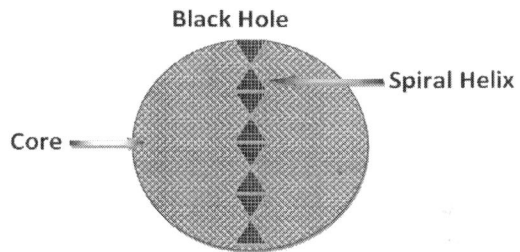

Review structure of a black hole. A central spiral helix at the core of the black hole pumps up and down while it rotates a core around itself made up of trillions of compartments.

Now, a massive black hole which didn't desire to grow into a sun, but rather maintain its position as the dominant massive black hole at the center of a galaxy, had to develop a new skill. To attain this new skill, the massive black hole had to make (evolve) new structural changes. Till then, every cycle the black hole had always added to the girth (breadth) and length of the central spiral helix axis operating its core.

As the black hole evolved further, each part of its spiral

helix expanded in length and breadth making the central helix a totally wider more hollow structure on the inside of its form. That helix took on so much extra breadth that a vacant area running the length of the helix developed, similar to a hollow pipe. Eventually, this looseness in the helix prevented it from spinning as much light into matter. The helix finally widened so much that it evolved into more of a conduit where light can pass right through. Basically, now the massive black hole only spun light into matter to continually build its internal structures, while never again adding more matter to build a bodily sun around itself.

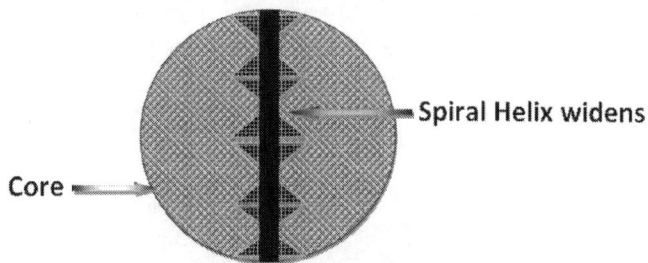

Massive (Supermassive) black hole with a wide hollow expanded area along its spiral helix.

With the development this new skill, the massive black hole was able to spend more time and focus on continually growing its internal structures, thereby totally moving faster than any other competitive black holes, enroute to growing into a supermassive size. This new strategy also allowed the massive black hole to pull in extra tons of light, using only some light for building internal structures, while passing the rest of the pulled in light right through its hollow helix and then spewing it out its poles.

In essence, the massive black hole at the galaxy center of had learned to breathe, attracting tons of light into itself and sending (pluming) much of that light outwards via its poles.

As the massive black hole learns to breathe, spew out and jettison light in plumes from its poles, it never has to grow a huge sun body and eventually have to go Supernova. It has strategized a method whereby it can live a much longer cycle than just 15 billion years and maintain its dominant position as a supermassive at the galactic centre of a galaxy.

Plumes of light are forced out the top and bottom of the super-massive black hole, as its helix becomes hollow. The pumping action of the supermassive black hole pushes the loose light from the inner hollow out the top/bottom of the spinning hole.

That was quite the feat for a massive black hole to be able to change its structure and the rules or laws pertaining to it, finally living right through the 15 billion year rule.

A Supermassive centering a galaxy grows much older.

Some supermassive dominant black holes have been at the center of their galaxy for up to 800 billion years. They're likely destined to remain in that dominant galaxy position for ever, never growing into a sun or going Supernova.

The supermassive black hole, using super spin, controls all the surrounding suns by tugging hard on their contents. However, the supermassive black hole is keeping down its own mass by expelling incoming light through its poles. The changes in function allow for wins on several fronts:

Firstly, the supermassive black hole doesn't have to grow into a sun. It can live long, control its galaxy, and prosper.

Secondly, it can spin fast because it stays as a black hole, not taking on the sluggish rotational slowness of a sun.

Thirdly, it can take huge amounts of light from the galaxy stars, then expel, spew and jettison that light as plumes out through its poles providing fresh light to the gigantic galaxy.

Fourthly, it can use the rotational powers and mass of the stars it pulls close to assist in the spinning of its spiral arms.

So for a supermassive black hole, humongous makes a difference. The rich get richer as the supermassive black hole of a galaxy can live on through all the 15 billion year recycles of the stars in its galaxy. By not having to grow into a sun, the supermassive black hole fends off having to recycle itself and instead it can maintain its stature as the hub of the gigantic galaxy for hundreds of billions of years.

Thus, a galaxy can grow ever larger. In the next 15 billion cycle of our cosmos (cycle number 68), our universe size will double. Within known galaxies, solar systems will become larger as more spheres are added to the systems. More solar systems will be counted as some existing ones will split apart during their Obliteration forming two solar systems instead of one. The star count within all the galaxies will double from 73 quintillion stars in our present 67th cycle to 146 quintillion stars in the 68th next cycle.

Clearly seeing a galaxy's supermassive black hole is not always easy. A bright halo surrounds the supermassive.

At the center of a spiral galaxy, a supermassive black hole bulge holds the galaxy together in a linear pancake formation. This supermassive black hole is surrounded by many large stars. Light plumes out the top and bottom of the black hole.

A supermassive black hole transforms to the point where it can attract more light than ever before. This strategy also allows the supermassive black hole to pull dominant stars into a cluster around itself, eventually unspinning those stars and gobbling up their light. Thereby, never allowing those stars to challenge for supremacy of the galaxy. Although the mass of the stars assist the supermassive black hole with the gravitational pull necessary to hold the galaxy together. In essence, the ingenious strategy of the supermassive black hole uses its prowess to make these stars into servants to assist in the controlling of the humongous galaxy.

These pulled-in stars glow brilliantly like a halo as there is a plethora of stars in concentrated numbers surrounding the supermassive black hole. Note: This halo makes it difficult to get a pure look at the supermassive black hole.

The supermassive black hole's pluming out of light also fits into its strategy. This galaxy might be far from the outer ocean of static light, so the supermassive strategizes closer sources of light making its galaxy grow and ensuring that its solar systems recycle on cue. It speeds this process by pulling large stars in towards itself and sending their light out for resident black holes to use in building solar systems.

However, there is always more and more light needed to run the operation of a galaxy. Thus, galaxies are continually attracted outwards towards the perimeter of our cosmos to be closer to the ocean of light where new light might come available. That is the scene where many new naked black holes are busy at work breaking off chunks of the static light, trying to consume it. Some of that un-captured and escaping light becomes available to current galaxies.

Therefore, galaxies are seen as receding outwards in our cosmos. Whereas, the innermost part of our universe is just empty space leftover as a byproduct after building occurred.

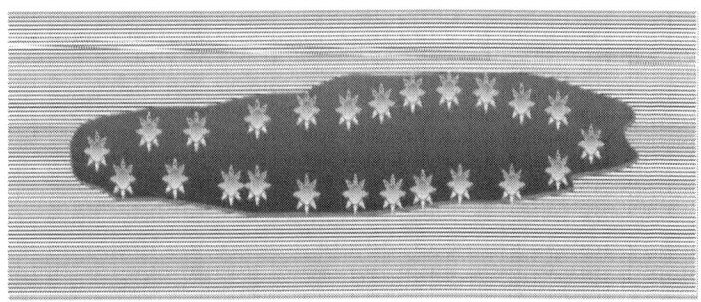

The billions of galaxies in our cosmos mostly travel outwards expanding away from the center, pursuing the abundance of light possibly available in the outer static ocean of light surrounding the perimeter of created space.

Is perfection within our cosmos ever 100% possible?

No. Not every piece of debris, whether comets, asteroids, dust clouds, gas or clouds, can be perfectly re-assimilated by naked black holes right after Supernovae or Obliterations occur in the solar systems and galaxies of our cosmos. Some of this unaccountable stuff, like the gaseous clouds, might never be assimilated properly. This is the 1% which will be studied for decades after we figure out the more important and more systematically relevant 99%.

Does this make our universe too imperfect or way too helter skelter? No. But, perfection to 100% recycle is likely never truly possible. There is always bound to be untamed debris floating around space or inside of solar systems, galaxies, nebula, gaseous clouds, and so on.

CHAPTER 11
TWO DIFFERENT
GRAVITIES

Understanding gravity. Okay, really, what is it?

If asked, a person would say that gravity is a force which holds us on our planet's surface. Gravity holds our moon in orbit around Earth. And our entire solar system of planets is held in orbit by the gravity mass of our powerful sun. All this gravity we take for granted. But, that doesn't really explain what causes gravity to be a force in our physical universe.

Trillion Theory says, 'Gravity is the strong gravitational pull which an ultra-fast spinning black hole extends outside of its core when that black hole is consuming light.'

Trillion also says, 'Gravity is the strong gravitational pull which a black hole extends to the surface and out past the surface of the sphere which it inhabits the centre of.'

So, which one is it? Trillion Theory says that it's both, depending upon the stage of a black hole. Overall, Trillion Theory advocates that, 'Gravity is a property owned by the black holes of our cosmos, and it can have different effects. Initially, gravity can bring things into a naked black hole. Thereafter, gravity can stop things from leaving.'

Using Earth as an example, the gravity owned by the black hole at Earth's core originally spun at an ultra-fast spin rate as an empty naked black hole when it first attracted and spun light into matter to fill its bowels and build a body of matter to form Earth. Now, this same black hole existing at Earth's core, spins slower since it has been slowed by all its acquired mass, yet it still extends its gravitational pull to hold us on the surface and also hold our moon in orbit.

Largest black hole of a solar system controls the system.

The size of a black hole determines the power and reach of its gravity. Gravity follows specific rules: larger black holes have a stronger gravity and can extend their gravity further afield. This means that a larger black hole sphere can trump and hold a smaller nearby black hole sphere in its orbit.

Example: The XL black hole at the core of our massive sun forces all planets in our solar system with smaller black holes at their cores into orbit around it. And since our sun's spin on its axis is counterclockwise, all spheres in our sun's Event Horizon revolve counterclockwise in orbit around our sun.

The black hole at the core of our sun spins in a counterclockwise direction on its axis, so every planet in orbit in the sun's Event Horizon revolves in a counterclockwise direction around our sun.

One black hole's gravity can trump another's.

Moons tag along with the planet they belong to as that planet revolves counterclockwise around our sun. But, the main orbit for a moon is the one it takes around its planet. The reason this is so is that proximity matters. Which gravity is closest to a moon? For a moon, its planet's gravity is the nearest gravity affecting it, so the moon revolves around the planet. Thus, the planet's gravity trumps the further away, yet stronger gravity of the sun.

Two different gravities: A black hole's gravity pull can guzzle light or hold matter in place.

The gravity of a black hole works in two different ways, depending totally upon a black hole's fullness.

Gravity for a naked black hole at the CREATION of matter phase works different than the gravity of a black hole inside the core of a planet or star during a LOCKUP holding phase.

Both these gravities have tremendous force, yet they act differently. The reason is that the black hole changes its duty as the black hole fills with matter.

Guzzling Gravity – starting from empty.

With a naked empty black hole, its gravitational vortex is dedicated strictly to sucking in light. Guzzling is the name of this process. A black hole is constructed such that its body spins around a central spiral helix which lengthens and shortens with each superfast powerful spin to perpetuate and accentuate a strong magnetic field. The black hole will use the power of guzzling gravity to attract and spin straight line light into matter during this CREATION of matter phase.

The gravitational force of the fast spinning black hole entices nearby rays of light. Rays are curved and slowed, then bent and spun acutely to the extent that the head of the light ray turns sideways and spins like a barber's pole thus creating a spinning helix which pulls and spins the rest of the tail of the light ray in behind to form an atom.

During the CREATION phase, light attracted by a black hole is slowed, bent and spun into matter.

Gradually, the Guzzling Gravity of the black hole attracts evermore light which is devoured as the hungry black hole jails more light into atoms of matter. The earliest formed atoms align with the black hole's helix to form a more powerful axis. As matter builds up, both the helix axis and the core body of the black hole increase in size.

This strong Guzzling Gravity during the CREATION phase has bazillions of tentacles leading out from compartments of the core bowels of the black hole. At the beginning of the CREATION phase, every single compartment is dedicated to guzzling, capturing and spinning light into atoms. But, once a tentacle compartment captures a ray of light, it has done its quota, meaning it can no longer perform guzzling. It must now dedicate itself to holding and LOCKUP. And, as more compartments fill they switch their dedication away from Guzzling Gravity to Holding Gravity.

Holding Gravity of a full black hole.

There comes a tipping point for the full black hole as its compartments become 100% dedicated to holding during LOCKUP. Thus, no further guzzling can occur. This lockup is so powerful that the black hole's Holding Gravity on its atoms extends out past the black hole's core to its surface to hold anything and everything from departing the surface. and flying out into space. For any living organisms on the surface there is difficulty moving away from the surface as all efforts to jump free or fly away result in a gravitational drop back down to the surface. It takes a tremendously strong effort, such as a space shuttle rocket, to escape the powerful hold of LOCKUP during Holding Gravity exerted by a black hole at the center of a sphere.

Now, suppose that the black hole at Earth's core is still attempting in vain to use its gravity to attract and spin more light into matter. But it can't as there is too much matter in the way. Us, being held on the surface and not flying off is perhaps just an accidental benefit of this attempt.

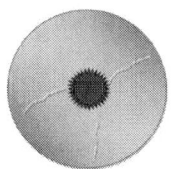

As the Black Hole grows larger, all tentacle compartments have captured sufficient rays of light to form a sphere (a moon or planet). Now, each of the compartments rededicate themselves to a new job, namely to Holding Gravity in order to hold spun atoms in place and prevent them from escaping.

For any planet, power of this gravitational hold extends out past its own surface into space to nearby moons, held in orbit. For a sun, this powerful gravitational holding extends far out to the planets.

The rest of the life of a sphere at this full stage will be dedicated to Holding Gravity, maintaining a strong LOCKUP of all of its atoms for as long as possible. That effort usually pays dividends for billions of years during a LOCKUP phase.

LOCKUP in the body of the black hole, now turned planet, can last for billions of years. However, after billions of years of LOCKUP, as the sphere further ages, the compartments of the sphere tire and begin to lose control of their contents.

The first freed atoms can form into extreme heat at the core of the planet. Further along in time, a few billion years later, the planet can even swell to the size of a small star.

Even further along after another billion years, the planet can lose control making its surface hot and liquid. Near the end of the LOCKUP phase, it can heat up and emit light as a sun. In the end, the sun can entirely fire up and finally explode as a Supernova. All light trapped escapes back to freedom as straight line light (ESCAPE), incredibly somehow escaping from the powerful gravity of a tired sun.

How light is able to escape our sun's powerful gravity.

It's puzzling how light can escape from our sun which possesses the strongest gravity in our solar system. This answer is tricky. We go back to the start of that solar system, and we give our sun a name - Henry. At our solar system's start, only naked black holes remained after a Supernova destroyed the old system. As the largest naked black hole of that new system, Henry attracted and spun the most light into matter. Henry was also the first black hole to gobble sufficient light to grow into an x-large planet, completely switching from guzzling to holding gravity. Possessing the most mass, Henry's stronger gravity kept smaller spheres in his orbit. So, Henry became the center sphere of our new solar system, a solar system which still lacked a sun. But, Henry was also the first sphere to experience a loosening of its atoms because it had been a total glutton and overeaten. Other spheres held in Henry's orbit were still guzzling. These black holes further sped the loosening of Henry's atoms. This loosening became rampant and Henry's surface fired up entirely turning him into a sun.

Thereafter, light rays freely left Henry's surface. The other black holes orbiting him soon ate so much of his light that they grew into moons and planets, entering phases of holding gravity. So, while sun Henry possessed the mass to hold the spheres in orbit, he lacked the necessary Holding Gravity to keep his atoms from unspinning back to original straight line light and escaping from his sunny surface.

Guzzling Time compared to Holding Time.

The amount of time a black hole spends spinning light into matter is short compared to the amount of time it will spend holding spun atoms in place.

Guzzling Gravity time duration (CREATION phase) may only be a few years, dependent upon the plentitude of the supply of light available to be spun into matter and also dependent upon the size of competitive black holes nearby.

Holding Gravity time duration (LOCKUP phase) may take anywhere from eons to up to fifteen billion years for a black hole which goes thought the entire cycle from naked black hole, to solid sphere, to large solid sphere, to liquid surface sphere, to small sun, to a larger sun, to Supernova death.

Several factors determine this eventual ESCAPE of light.

♦ A black hole always overeats during CREATION phase. It is always too greedy in its desire to devour light.

♦ Speed of spin of a black hole gradually slows over billions of years as it goes from a naked black hole to full sphere.

♦ Light trapped and spun as a ball inside an atom of matter always wants to escape back to free straight line light.

♦ Light will always be successful in its escape from its jail in matter, even if it takes eons or billions of years.

♦ Light deep within the central core of the black hole of a sphere will always loosen the soonest. This is the area of the oldest spun matter.

♦ Deep in the core, the first escape of light and heat begins, and that's where a fire starts up.

♦ This internal fire deep in the core always tussles its way outwards to the surface as the heat expands.

♦ This loosening and resultant fire always moves outwards from the core through fissures. This fire finally reaches the surface as molten lava. Billions of years later on, this fire can become so rampant that it can liquefy the surface of the planet, making it into a fiery sun.

Speed of spin changes with the type of gravity.

During Guzzling Gravity (during the CREATION phase) the naked black hole demonstrates incredible spinning speed used to devour light and spin it onto matter.

This incredible spin speed can even affect other spheres nearby as there can be an extended draft out into space. Moons and planets in the direct vicinity of that draft are susceptible to experiencing a premature loosening of the atoms comprising their bodies. Nearby suns can experience a premature loosening of their atoms from an attack by a black hole causing a sun into an early Supernova.

During Holding Gravity (LOCKUP phase) the spin speed of a black hole slows dramatically. That lower spin rate is best suited for holding. Thus, as the mass of the sphere increases, so does the outward extension of its Holding Gravity away from the sphere to hold moons or planets in orbit.

During the last of the holding gravity (ESCAPE phase), when light is escaping from the surface of a sun, the rate of spin for the black hole at the sun's core slows even more.

Supernova affects the gravity of all involved spheres.

At the moment of any Supernova, the direction of spin reverses for the black hole at the core of the sun. This is caused by the whipping backlash from the instant release and escape of tons of atoms (for every action there is an equal and opposite reaction). In the end, this backlash of speed splits the axis core of the black hole into two parts.

The planets and moons of the solar system are released from the gravity hold of the Supernova sun. Spheres closest are obliterated (they have their own quick expansion and supernova as their atoms all burst into light). Some distant spheres of the solar system may survive and be driven from the blast area further out into space as solitary rogues.

It is during this reproductive phase where newly twined naked black holes receive the direction of rotation on their axis and the direction of the gravity Event Horizon which will be with them for the entire cycle of that particular black hole and for the sphere which will grow around that black hole.

Conclusion to the two types of gravity.

So in this chapter we've seen that there is the gravity of a naked black hole (uncloaked black hole) which attracts and pulls light into the black hole to be spun into matter. And, there is the later gravity of a full black hole (cloaked black hole) where the black hole is hidden away from sight at the core of the sphere which it spun around itself.

In our own solar system, and generally across our entire cosmos, black holes are cloaked away hiding away inside of full spheres. Their gravity during these long episodes is designed for holding atoms of matter, keeping the body of the sphere intact, holding objects on the surface from flying off into space, and for holding other spheres in orbit.

Not till Trillion Theory are we seeing what gravity really is and what causes it - namely black holes either naked or cloaked inside of a sphere. Till now, astronomers have seen black holes gobbling up light, but they had never discovered that the actual true purpose of black holes was as engines to build spheres and engines to control spheres.

Now in Trillion Theory, new terms *Uncloaked Naked Black Holes* and *Cloaked Black Holes* become important terms when fully understanding the gravity prevalent throughout the solar systems and galaxies of our universe.

As we discover more and more about black holes and how they built (and still build) spheres, solar systems, and galaxies we will no doubt discover more about gravity. So far scientists have never been able to really manipulate (increase or lessen) or control gravity on a large scale for any good purpose, perhaps someday they will.

CHAPTER 12
SUPERNOVA
AND
OBLITERATION

Trillion Theory speaks a lot about the Supernova of a sun, and the resulting Obliteration of its solar system. After those two events, comes Replication by all the surviving naked black holes which were at the core of each and every sphere in the old destroyed solar system. This is followed by the Reformation re-building by these naked black holes of a brand new solar system in that particular area of a galaxy. Lastly is the extremely long holding phase for the spheres (the situation of our present solar system is an example).

This entire cycle destined to last about 15 billion years; with most of the 15 billion years dedicated to the holding phase as a cooperative generated between the spheres. Whereas, Supernova, Obliteration and Reformation take only short periods of time. Here is a rough timeline of one cycle.

Supernova > (explosion is a short time episode)
Obliteration >> (solar system spheres destroyed)
Replication > (easily the shortest phase)
Reformation >>>>> (solar system rebuilt)
Holding inside of a sphere (phase of billions of years)
(the longest lasting phase and most visible to the eye)
>>>
>>>
>>>
>>>
Onto the next Supernova > (the next cycle begins).

Of Supernova, Obliteration, Replication, Reformation, and Holding, each is equally important to the recycling process.

In this chapter and the next, Trillion Theory more closely examines and determines the specifics of how and even more importantly why each of these phases is occurring.

Supernova - End of the cycle for an old star.

It is an incredible brilliant sight to see a star explode. While many of them happen across the vast expanse of our cosmos each day, you have to be in the right spot at the right time to catch a Supernova explosion in action.

Supernova of a star happens for two reasons.

Scenario one occurs most times. The sun reached old age. It has achieved over a 13 billion year age in its current cycle. It experiences extreme difficulty in controlling its contents as an inordinate number of its atoms are unraveling and escaping away, back to freedom, as straight line light. As this process escalates, more and more atoms unravel at a hectic pace. Larger solar flares frequently burst from the surface. The star expands, providing an over-abundance of heat and light to its solar system. Sun expands enormously engulfing the very closest spheres held in orbit. Finally, the sun loses total control of its contents and explodes into a Supernova.

Scenario two is quicker: A naked black hole comes in close proximity to the sun, so the exact same changes occur as in scenario one, only at a much accelerated pace. The powerful naked black hole eats ferociously of the sun's light causing the sun's contents to unravel even faster.

Right at the precise moment of Supernova, the escalations mentioned above turn totally violent as the star explodes. At that exact moment of Supernova's outward explosion, all its contents are flung out into its solar system. The sun's atoms instantly unspin and fire outward as freed straight line light. The sole survivor is the black hole at the sun's core.

Obliteration – end of the cycle for an old solar system.

The brilliance of a Supernova by a single star is enhanced by the explosive Obliteration of most of its solar system. The explosion's outburst wave from the sun engulfs most of that solar system. Every nearby planet and its moons in that solar system quickly swell and melt, flash-evolving from the heat's intensity. They too lose control, unspinning many septillions of atoms back to light. For any planet to survive the long reach of the Supernova, it would have to be several billions miles away in a distant orbit. Unfortunately, those closer are not so lucky. The destroyed moons/planets never get to see a longer existence, or an opportunity to further evolve.

Black holes survive Supernova and Obliteration.

The only part of the sun which survives the Supernova is the black hole at its core. A black hole cannot be destroyed.

The only part of each and every planet and moon which survives Obliteration is the black hole which was at its core. A black hole cannot be destroyed.

In that graveyard of the old solar system, after Supernova is followed by Obliteration, there no longer is a sun or any planets or moons, except for perhaps a distant planet lucky enough to be in an outer orbit far enough from the blast to survive. Even so, it may be flung further out into space.

So, what is left after Supernova and Obliteration?

There is little available evidence that the old solar system ever existed. All light which had been trapped for billions of years as matter in the spheres is now flashing outwards attempting a bid to escape the area. Or so that light thinks.

At this point, it is difficult even for the most powerful telescope to hone in and see clearly what has occurred. For unknown until Trillion Theory, in the aftermath graveyard, all the black holes have survived and they now are uncloaked, naked, and hungry, eager to once again devour light.

CHAPTER 13
REPLICATION
AND REFORMATION

Replication is a term used for the reproduction of one black hole into two new black holes.

Reformation is a term used as for the re-building of a new solar system by naked black holes after their old solar system had been destroyed by the Supernova of its sun.

Replication. Trillion Theory states that Replication by black holes grew the cosmos over the past trillion years.

This is one of the parts of Trillion Theory which is most amazing. How naked black hole can replicate themselves (splitting into two twins) after the Obliteration of their solar system and the destruction of the matter which used to comprise the body of their spheres.

Black holes are survivors, able to survive anything, and able to replicate themselves during the traumatic events of a Supernova and Obliteration.

Trillion Theory uses both the terms Replication (twinning) and Reproduction (making two new black holes by splitting the old black hole, a fission reproduction) in describing how black holes increase their numbers. This Replication feature of black holes is what has allowed our cosmos to start small at origin a trillion years ago and then to grow exponentially with every new cycle to its present size of 73 quintillion stars within billions of solar systems within multitudes of galaxies.

During Supernova, the powerful explosion of the matter of a sphere in an instant accelerates its spin. For every action this is an equal and opposite reaction, as the black hole at the core backlashes at that instant, splitting in two.

Spiral helix is an interior component running
top to bottom inside of a black hole and is
responsible for spinning the black hole on its axis.

The helix splices and splits when a Supernova exacts
its force on the black hole which was residing at the
sphere's interior. This action splits the entire black hole.

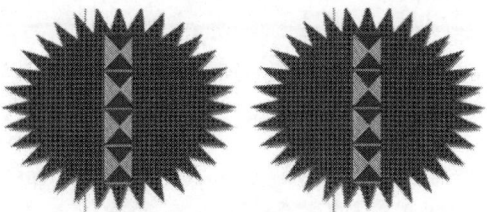

The one black hole splits and duplicates during Replication
to form two new naked black holes. The force of this split
drives the two new black holes away from each other.

Replication (doubling process).

When the sun violently explodes as a Supernova, the only
thing which survives the blast is the indestructible black hole
at its core. At Supernova, the sun jumps into an instant fast
rotation, as the backlash from tons of matter leaving its
surface spins it ferociously like a top.

There is an equally violent reaction within the core, causing the emptied black hole helix to split, replicating into two twin black holes.

Initially, these two identical naked black holes are thinner than the previous one. But, they will quickly get back to being even larger as they gobble up light to increase the length and girth of their spiral axis and also expand and increase the compartment numbers comprising their body.

While the emancipated body of the sun returned to straight line light, the black hole remnant at its center core could not be destroyed as it survives and spits in two. The result is two new naked binary (twin) black holes replacing the previous one. They can spin in the same or in opposite directions to each other. This feature explains why spheres of our universe can spin in either direction as determined by direction of spin of the black hole at their core.

The same phenomenal splitting and doubling transition occurs at the axis of every black hole at the core of every obliterated planet and moon in the solar system. In this fantastic rebirthing event, with the twinning asexual fission replication of all the naked black holes in the vicinity of the previous solar system, there is now double the number of black holes as compared to the old solar system. These newborn naked black holes (now double in number) are poised to double the star/planet/moon population of the new system of the region. The new next fifteen billion year cycle of this part of the universe is ready to restart, even bigger than before.

Reformation. The building of a new solar system by the surviving naked black holes.

Every solar system rebirth is as different and unique as a snow flake. The split black holes can be flung in a multitude of directions and distances from one another. Sometimes new binary black holes remain in close proximity to each other; other times they are scattered far apart. Therein, one larger double-the-size new solar system could begin to unfold; or two or more new smaller far-apart solar systems.

Especially with larger solar systems, the immense power of Supernova followed by Obliteration normally causes more than one solar system to form during the Reformation.

Definite progression of events occurs after a Supernova and Obliteration.

• Double the number of replicated naked black holes results.

• One bigger group of naked black holes could result, or two or three flung apart groups. Either way the number of black holes in the cosmos increased.

• Each new naked black hole is immediately competitive with each and every one of the other naked black holes in the area. It is survival of the fittest as the early birds (the two black holes resulting from the old sun) are larger and poised to win the battle for light in their particular area. That area is without a new sun as yet, and the moons/planets are just beginning to form as the black holes spin matter around their bodies. Gravitational forces see larger black holes holding smaller holes in orbit, setting an early configuration.

• Some light does manage to escape from the vicinity right after Supernova and Obliteration. This escaping light exits out into space, travelling freely, after billions of years of being jailed inside of matter.

• The rich get richer. The largest naked black holes have the upper-hand on all the lesser sized black holes. Brilliance occurs when some of the light can't escape and gushes back inwards towards the largest black hole. This is prolifically seen in one of the black holes where the old star existed, as that large early bird is ready to recapture any of the light exploded away from the moons/planets during Obliteration.

(We can feel sorry for light which spent billions of years in entrapment only to be quickly recaptured).

Reformation. Fight for light in building new spheres.

Right after Supernova/Obliteration, the region of space around all the new naked black holes becomes a war zone. All the holes spin at their mightiest speed to cannibalistically snap stands of light to spin into the depths of their core compartments. This battle for light rages with now double the number of black holes. Note: The number of black holes after Obliteration is double, but they might now exist in one solar system setting or two separated systems.

In the new system, temporarily comprised of just naked black holes battling each other to consume the most light, the old graveyard in space neither remembers nor shows anything of the old solar system of sun, planets and moons, which got wiped out. All that remains from the old solar system is double the number of naked black holes and light flung outward from the Supernova and Obliteration. There is little true evidence in the new graveyard that the old solar system ever existed, for sphere recycling is very efficient.

The two XL black holes which split from the center of the old sun have a distinct advantage having done Replication before any lesser black holes at the cores of the moons or planets. An XL scoops up light from those smaller neighbors when they explode during Obliteration.

XL has the upper hand to become even larger compared to the lesser black holes, as it spins the most light to add to the mass necessary to hold all lesser black holes in orbits. This battle for light will decide the new head honcho of the new solar system. XL gobbles the most light, gluttonously filling up, becoming the first of the black holes to complete a body. XL will be the first sphere to change from Guzzling Gravity (light capture) to Holding Gravity (atoms controlled).

Thus, this XL is the top candidate to have its atoms loosened and one day and evolve into the solar system's new sun.

Of course, each new solar system is totally unique, no two snowflakes exactly alike. With each new solar system; each planet/moon has its own size, mass, composition, climate, surface activity, orbit, and speed of rotation on its axis.

However, how a solar system (or a galaxy) ends a certain way is more than just pure happenchance. Definite specific Universe Laws exist between black holes, suns, planets, and moons, providing similarities in organizational makeup.

Black holes, those building engines of the spheres of our universe, are guided by a set of specific Universe Laws when they form solar systems and then galaxies.

However, don't ever get the idea that naked black holes are always nice guys. They are absolute ravenous battlers when fighting against other black holes over the attraction and consumption of light - all done under the guidance of Universe Laws pertaining to black holes and spheres.

These Universe Laws handle both 'rules of co-operation' and 'rules of battle.' All the spheres in solar systems find times when they work with one another and other times when they are totally egotistical in battle during the building of a solar system. These battling scenes are akin to pigs at a trough; a contest where the strongest-fastest wins the order of finish. Nonetheless, Universe Laws are obeyed. Study any solar system and it will abide by similar Universe Laws.

For, every black hole wants to do its best; an innate desire to be Queen Bee. In solar system settings, they battle for control and the winner becomes the sun. They always want to be the biggest controllers taking charge of other spheres in orbit. Every XL massive black hole desires to someday control an entire galaxy as a supermassive black hole.

More about Cosmic Universe Laws in the next chapter.

CHAPTER 14
LAWS FOR
BLACK HOLES
AND SPHERES

Inside of Trillion Theory there are Cosmic Laws for:
▪ naked black holes of small, large, x-large, massive and supermassive size.
▪ spheres of small, large, X-large, and massive size.

It is important that these Universe Laws be categorized as these laws are the links which tell us why many similarities exist between certain types of spheres, and why there are similarities of organization with solar systems, and why there are similarities of organization with spiral galaxies.

These similarities result from consistent Universe Laws (rules of engagement in cosmic actions) which regulate the activities and relationships of black holes and also spheres. Classifying these Universe Laws begins with black holes and then proceeds to spheres, as follows:

General Universe Laws for naked black holes:

• A naked black hole has an insurmountable desire (mission) to devour light, and a directive to spin that light into atoms of matter in for the purpose of building a sphere.

• A naked black hole can devour light in amounts relative to its naked empty size. And, a naked black hole always wants to increase its size from one cycle to the next.

• An empty black hole always eats light faster and in greater amounts than a full black hole such as a planet.

• A naked black hole, because of its spinning power, creates gravity extending strongest outwards from its equator.

Universe Laws for SMALL naked black holes:
- A small naked black hole devours less light to spin into matter around itself than a larger naked black hole.
- A small black hole exerts less gravitational pull than larger black holes.
- A small naked black hole has difficulty winning battles for light or using its gravitational pull against larger opponents.
- A small naked black hole spins only tight atoms around itself, thereby building a hard surface.
- A small naked black hole spins a small sphere of matter around itself. Therein, a small black hole, once full, would be at the center of a moon or small planet in a solar system.
- A small naked black hole would spin too small of a sphere around itself to ever have a chance of evolving into the sun of a solar system.

Universe Laws for LARGE naked black holes:
- A large naked black hole can devour more light to spin into matter around itself, than a smaller naked black hole.
- A large black hole will overpower smaller black holes in the battle to spin light into matter.
- A large naked black hole exerts a stronger gravity pull than a smaller naked black hole.
- A large naked black hole's gravitational power reach can hold smaller spheres in orbit.
- A large black hole spins looser atoms, making it the fitting center for a larger sphere such as a large planet.
- A black hole of larger size has better odds of someday evolving from a moon to a planet to a star.

Universe Laws for EXTRA LARGE (XL) naked black holes:
- An XL naked black hole can devour more light to spin into matter than a small, medium, or large naked black hole.
- An XL naked black hole will overpower smaller black holes when it comes to devouring light to spin into matter.

- An XL naked black hole exerts a much stronger gravity pull around itself.
- An XL black hole, because of its greater gravitational reach, can hold small and large spheres in orbit around itself.
- An XL black hole spins the loosest atoms as a body, which makes its surface more gaseous and less solid.
- An XL naked black hole has top odds of further loosening and evolving into the sun of a solar system.

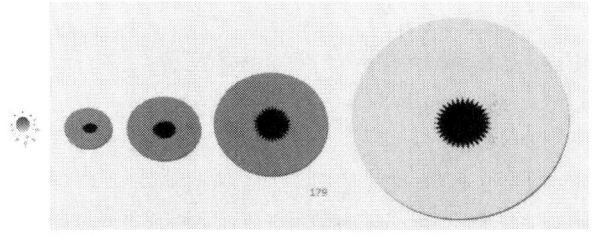

Black holes have varying sizes regardless of whether they are naked or if they already have a body of matter spun around them. Left to right above: naked; small at center of small moon; medium at center of a medium sized solid planet; large at center of large gas giant planet; XL at center of a star such as our sun.

Universe Laws for MASSIVE black holes:

- A massive black hole possesses a superior size and gravity advantage of all the lesser black holes in its area.
- A massive black hole uses this advantage in the building of an x-large solar system around itself.
- A massive black hole builds a massive sun around itself.
- A massive black hole inside of a massive sun has superior gravitational mass to also hold other solar systems in orbit forming a small galaxy around itself.
- A massive black hole inside a massive sun will tire of going Supernova and have to rebuild its small galaxy.
- A massive black hole has a goal to become supermassive and one day become the center of a gigantic galaxy.

Universe Laws for SUPERMASSIVE black holes:
(supermassive black hole at the center of a galaxy).

▪ A massive black hole's top priority is to become the center of a galaxy, therein becoming a supermassive black hole.

▪ a supermassive black hole's goal is to control its galaxy for a period of up to hundreds of billions of years, thereby overcoming the normal 15 billion years allotted for a normal solar system or a tiny galaxy.

▪ a supermassive black hole deploys new strategies with its structure in order to break the rule of 15 billion years.

▪ a supermassive black hole deploys new strategies allowing itself to live much older, up to 500 billion years or more.

▪ a supermassive's structural changes allow it to remain as a supermassive black hole and not have to grow into a sun and someday go Supernova destroying its galaxy.

▪ a supermassive's black hole doesn't need to take on the sluggish rotation which would occur if it became a sun.

▪ a supermassive black hole's structural changes allow it to pull in tons of light, therein able to hold large suns around itself (seen as a halo), and pass this light right through its hollow helix pluming and jettisoning out its poles.

▪ a supermassive's black hole utilizes the gravitational mass of the suns which it holds in a cluster around itself to assist with the holding and spinning of the entire gigantic galaxy.

The supermassive black hole at the hub of a spiral galaxy
is surrounded by a halo of stars. The tons of light pulled
inward by the supermasive plume and jettison
out the top and bottom of the black hole.

This concludes the Universe Laws for naked black holes. Now we move onto the Cosmic Laws pertaining to spheres; these laws deal with the situations occurring after black holes have built spheres around themselves.

General Cosmic Universe Laws for spheres:

▪ A sphere with a full black hole hidden and cloaked in its core has all of its contents held in place by the spinning gravitational force of the black hole at its core.

▪ A sphere exerts gravitational pull around itself relative to the size of the black hole at its core and the amount of the mass of the sphere.

▪ A sphere with a larger black hole at its core and more body mass extends its gravitational pull further outwards.

▪ A sphere with a black hole at its core extends the pull of its gravity strongest out along its equator.

▪ A sphere formed around a larger black hole can hold more spheres in its outer orbits; while smaller spheres are held in orbit by larger spheres. Therein, planets are held in orbit by a sun which is larger than them, and moons are held in orbit by larger planets.

Cosmic Universe Laws pertaining to sphere population:

▪ The black hole population and the star population of the universe doubles with each 15 billion year cycle.

▪ The black hole at the core of the central sun of each solar system gains in proportion to the smaller holes at the cores of planets and moons.

▪ Space grows ever larger with the billion to one ratio of space left behind after light is spun into matter.

▪ Expanse of the entire universe accelerates outward with each new solar system being pulled towards the direction of the static ocean of light on the outer perimeter.

▪ The old graveyard provides no fossil evidence to indicate that the old wiped out cycles of solar systems ever existed.

Universe Laws relating to SMALL SPHERE size (example: a small moon or planet in our solar system):
- Small spheres are controlled in orbit by larger spheres.
- Small spheres end up being moons of larger spheres.
- Small spheres are most easily pulled close to their sun.
- Small spheres have denser packed bodies, concentrated with tight element atoms. Their surfaces are harder.

Universe Laws relating to LARGE SPHERE size (example: a large planet in our solar system):
- Larger spheres evolve more than smaller opponents.
- Larger spheres have more gravity for the holding of other spheres in orbit.
- Larger spheres are able to stay further away from a sun.
- Larger spheres are less likely pulled close by a sun.
- Larger spheres have looser gaseous surfaces.
- Larger spheres in size-mass exert stronger gravity pull.
- Larger spheres have a greater possibility of evolving and growing into a star.
- Larger spheres (in far out solar system orbits) might survive a Supernova by their sun.

Universe Laws for EXTRA LARGE SPHERE size: (example: sun of our solar system; and stars).
- XL spheres have an extra-large black hole at their core.
- XL spheres easily evolve over small opponents.
- XL spheres have more gravity to hold spheres in orbit.
- XL spheres possess traits and power to become the central sun of a solar system.
- XL spheres, when they were being formed, had the loosest atoms of matter spun into their bodies.
- XL spheres generally have a liquid-gas body.
- XL spheres are first to lose control of contents.
- XL spheres (suns) can mushroom and explode.

CHAPTER 15
HOW SPACE
WAS INVENTED

Nearly everyone's perspective rationalizes space as just this emptiness (this box) which was already there, waiting for our universe spheres to take occupancy. This cold, empty, weightlessness space was supposedly already in place.

However, my Trillion Years Theory advocates that space was not there first. And this neo out-of-the-box type of thinking really throws people for a loop.

An easy experiment depicts space.

Spread out iron filings to completely cover a table. Think of the iron filings as the static ocean of light. Then, randomly toss 20 magnets into the table full of iron filings. Think of magnets as symbolizing black holes. Of course you will find that the iron filings are drawn towards and adhere to the 20 magnets. Think of iron filings stuck to the magnets as matter forming atoms on a moon or planet. Further examination shows what is left behind to be empty areas on the table where the iron filings used to be. Think of this emptiness as SPACE. Each magnet (moon, planet, sun) now weighs more with heavy iron filings adhered to it; the empty vacated barren table areas now weigh nothing, (like empty space).

Thus, Trillion Theory shows space as just a void empty byproduct left behind caused by the formation of spheres.

When black holes spin light into matter, all of the light's properties end up as spun atoms of matter. Space is just an empty left-behind byproduct. That's originally how space was created. Thereafter space was and is available as a super highway for light to return and travel through.

Definition of space.

Trillion Theory explains space as the empty nothingness byproduct left behind when light from the static ocean of light was spun into matter as spheres. Space is the empty footprint spun light left behind when spun into a sphere. Light basically laid its own return highway.

Trillion Theory puts it this way: first there was the static ocean of light; black holes were introduced into the ocean of light; these black holes took static light from the light ocean and spun that light into matter surrounding themselves; space was the empty byproduct left behind when static light from the ocean of light vacated its original area.

We can also state that space is that nothingness that the spheres of our universe seem to be sitting in. Space, that vast emptiness, takes 99.99999% of the room in our cosmos. Galaxies, stars, planets, moons, comets, meteors, asteroids, and gaseous clouds take up less than 1% of the room.

Space offers no resistance as light from stars travels for billions of miles and years effortlessly through it. Thus, space can best be summed up as the nothingness housing certain matter (spheres) and a highway permitting free passage. Space is empty, black, weightless, odorless, and frigidly cold (all the nothings that light left behind when it was spun into matter). Neo Trillion Theory here depicts how space actually came to be as a left-behind byproduct of light's departure.

In Trillion Theory, we can also think of space as the huge cavern created when black holes loosened and ate trillions of miles of light from the static ocean of light. Trillion Theory says that space is simply an after-the-fact empty byproduct derivative left behind as a return highway when light spun into matter. Space is weightless and frigid, while light is the carrier weight and heat and deposits those properties into the atoms into which it spins when forming matter.

Back to the origin of the universe.

All that existed at the origin of our universe was the static ocean of light. In its toolbox, this light carried products and byproducts. When the first black hole was introduced into the static ocean of light, that black hole freed static light from the ocean of light and spun that light into matter.

Two things immediately happened: firstly, light spun into matter which created the first sphere around a black hole; secondly, space was created and left behind as a footprint where light used to be. As more black holes populated our universe, space increased in size, expanding outward.

Occupying our universe was the endless static ocean of light. Then, a black hole was introduced into the static ocean of light. It devoured light spun into matter to form a sphere around itself. All properties, such as weight and heat carried in light's tool box, went with light into the matter of the body of the black hole. The large dark area between the black hole and the static ocean of light on the outside became empty space.

Free passage on express highway.

Light, which travels through the final frontier of space, from distant star to another part of our cosmos, is given this total free passage as space never charges any toll and never provides harassment. Thus, there exists this nice covenant between the passing light and space, the express highway provider. Space is the able servant of the universe, while the masters of our universe are light and black holes.

Light can travel through space for billions of miles and many light years. We have never seen how far light has been able to travel. If the universe is a trillion years old, some light has been traveling for that length of time. On the other hand, certain light may have gone through many cycles of spinning into matter and unspinning back to light.

At the outer edge of space, a person would see the endless static ocean of light, like a wall of stacked lumber. For all that existed prior to the formation of spheres was this solid ocean of light. As the universe expanded outwards it utilized more of this light leaving more interior empty space.

The shape of the space of our entire cosmos.

Trillion Theory shows that the general shape of the space of our entire cosmos is quite linear like a cucumber. Early in our universe, black holes could have been eating along a very linear plane. But, black holes can tilt away to more strategic angles from the linear plane when battling for light against other black holes. Therein, the space of our entire cosmos might be fairly linear overall, but not perfectly flat.

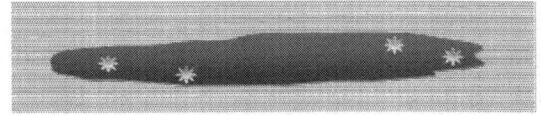

Black holes are shown eating light from the static ocean of light situated at the outer perimeter. These black holes could have been eating in any swiveling direction along a fairly linear plane. But, the black holes didn't have to stick to an exact axial tilt, thus, our universe in not perfectly linear. Nor is our universe anywhere close to spherical as should have been the case had Big Bang ever happened. But, it didn't.

Why space is so vast.

Black holes played the vital role in the determination the size of space. The distance between stars is so immense and space is so vast because of the disproportionate ratio which exists between the small sizes of the spheres compared to the billions of miles of space left behind as empty byproduct from light being spun into matter. This ratio is staggering.

A formula between matter and space.

A formula could be used to find a ratio between spheres and space. Why is space so vast between stars?

Space is much more famous for what it doesn't contain, than for what it does. For, space is an empty zone highway existing without weight, material, or warmth.

Even while there are billions of spheres and galaxies, empty space still occupies by far the greater area, a ratio heavily skewed: stars only occupy .0000000000000001%; while space takes up .9999999999999999% of the available room in our cosmos. Comparatively speaking, matter is so densely packed, that one atom of matter spun from light leaves one trillionth that amount of space behind.

1 atom of matter = 1,000,000,000,000 area of space. Such a formula finds difficulty with absolute accuracy. If we used our solar system as an example, we sum the volumes of every sphere and come up with an overall volume of the spheres in our solar system. Then, we take that volume and compare it to the volume of space left behind when our system formed. This is a most hair-raising calculation as it is 4.2 light years through space from our solar system to the nearest Centauri System. It would take an Earth spaceship over 50,000 years to travel to Centauri. So, the amount of space between here and there is gigantic, and how much space was also created after the formation of the Centauri system or our system is also a consideration.

Another more micro way to calculate the ratio of matter to space is by examining an atomic explosion. During such a nuclear explosion, the nucleus of a shattered atom releases tremendous pent up energy. This immense energy takes up a vast area in ratio to the spec of matter, a million, billion, or trillion to one. This nuclear release of energy (ESCAPE) is a mirror reverse of the spinning of matter (CREATION).

Space touches the static light ocean on the perimeter.

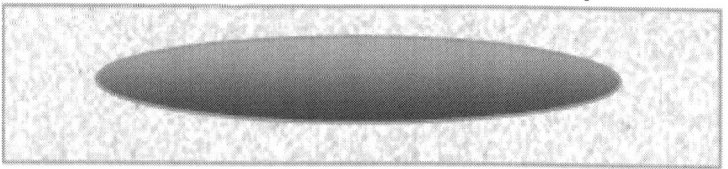

A reader might ask, "How is space so empty when light, from the outer ocean of light, could simply leave that ocean and flow inwards to fill up the space?

A clue came from a news article released in September of 2013. The article entitled 'Scientists Freeze Light' showed another of light's yet-undiscovered properties. You see, light has so many different speeds: speed in a straight line; a slower speed when bent; slower when passing through thicker mediums; slower again passing through diamonds; and slower again when spun into matter inside of an atom (this speed is dependent upon type of element atom).

Now, these scientists had discovered for the first time ever that light can actually be taken to a zero speed by freezing it. Scientists had frozen what is regarded as the fastest thing in our universe (that we know of at present). The feat was not anything easy like placing light into a freezer drawer. Rather, scientists converted light coherence into an atomic coherence as a crystal and shot laser light through and trapped a laser beam for a full minute at a standstill, halting light's fantastic speed.

Hurrah, this gave Trillion Theory more clout by showing why the static (frozen-like) ocean of light doesn't just flow light into empty created space left behind by the spinning of light into matter. The reason is that the ocean of light exists in a type of static frozen state.

Let's revisit the onset of our cosmos a trillion years ago when our cosmos began with the ocean of light. The ocean of light was filled with static light (not in motion), sort of just frozen in place. Yet, the word frozen here does not refer to the temperature 'frozen.' Rather it refers to a unique means of suspending light so that its speed is zero and it sits and waits at rest. The ocean of light was at a standstill.

This outer ocean of light remains frozen forever unless its light is pulled away by the powerful spin of a black hole.

Remember how the first black hole(s) at the origin of our universe attacked the static ocean of light, loosening and devouring it. Obviously, the tremendous spinning speed of a black hole loosened the light being held at a standstill. Once that light was jerked from the ocean, it immediately sped up to the speed of light. But, this freedom to fly was cut short for this light as it was quickly attracted, bent and spun into atoms of matter deep inside the belly of the black hole.

Once snapped, that spun light would have created some space left behind itself. In reality, the black hole would have eaten deeper into the static ocean of light creating an empty dark void cavern of space within the ocean.

Today, space has expanded to incredible limits as trillions of black holes have approached the light ocean, broken away and yanked stands of light from the ocean wall and devoured them. Light not pulled from the ocean of light has remained static at zero speed for the past trillion years. An inexhaustible supply is still ready and waiting to be utilized to build more and more spheres with our cosmos.

So, space is the empty byproduct which light laid-out as a future highway for itself to use whenever it escapes from a sphere. Various parts of the space highway has been utilized for hundreds of billions of years as a travel means for light rays, spheres, galaxies, comets, humans, and perhaps even aliens. The utterly huge space highway has expanded each time more light has been drawn from the static ocean.

CHAPTER 16
HOW TIME
WAS IMPORTED

'Time is an illusion.' Albert Einstein.

Many people have taken a stab at defining time, such as 'Time is an overseer of all things.' Trillion Theory presents a fresh new look at time and how it came to be a feature and a so-called overseer of our cosmos.

Time is a tricky subject. It marches on unabated. Billions of people on Earth get to share the ticking away of time each day. Of course we often get warped or skewed views of time. When we wait impatiently, time drags endlessly. Yet, at the end of a long highly competitive day, it is difficult to recall back to the morning. It seems like days have passed.

Trillion Theory definition of time is far different than ever proposed by anyone before. It includes: what time is; how it was imported to our universe; and also where it exists or doesn't exist inside of our cosmos.

Definition of time.

Trillion Theory definition of time: 'Time is imported inside of our universe right along with the spinning of light into matter. The measurement of time is the duration that light stays confined, spinning as matter below the speed of light.'

To elaborate: at the start of our universe, when there was only the static light ocean, time was nonexistent, but locked up inside of the waiting wall of light. Time was (is) a concept invented specifically for our universe, designed as a specific property inside of light. It was hidden away inside of the zillions of strands inside of the static ocean of light, waiting to be imported and become active in our universe.

Not until a black hole took light from the static ocean of light and spun it into matter did time begin in our universe. Importation occurred nearly one trillion years ago as the first matter was spun and our universe's clock began ticking. At the moment of that spin, time imported into our universe, carried here inside of light's tool box. Time was an imported derivative brought into our universe by light and let out of the genie's bottle when that light escaped the ocean of light and was spun into matter around a black hole sphere.

Think of a ray of light traveling through space. It moves at the speed of light where time stands still. Light carries in its tool box a quality known as 'time' which stays put until such time as that ray of light slows below the speed of light. Once slowed, a ray of light sees 'time' creep from its tool box. And once light is spun into matter, the time property is active in that particular area.

And once a ray of light gets spun into matter by a black hole, trapped inside of an atom of matter where it spins for eons below the full speed of light, time will exist around that atom until that light escapes and returns to the full speed of straight line light. And at that full speed of light, time goes back into lights toolbox, seeming to once again stand still.

How time is measured.

At the full speed of light, time does not tick, basically standing still. It was Einstein who first showed this feature of the speed of light. If a spacecraft left Earth at the full speed of light, the people on the craft wouldn't age at all living in a no-time zone; whereas people on Earth would age in a zone which operates well below the speed of light.

Therein, time is calculated as the duration during which light moves below the full speed of light.

The starting point for time.

At the point just before the beginning of the building of

our cosmos, *even before time had been let of the genie's bottle*, all that existed within our physical cosmos was the static ocean of light extending endlessly in every possible direction. At that point, time did not exist in the ocean of light. However, every single static light ray within the ocean held in its tool box the ability to carry the feature of time along with it should that be released from the ocean of light.

With the introduction of the one initial naked black hole into the ocean of light, it did the duty it was programmed for. Deploying incredible spinning speed, the naked black hole broke chunks of light free from the static ocean of light, attracted it, pulled it in, and then spun the light into matter.

This breaking light away from the ocean of light was the beginning point for time in our cosmos.

A singular naked black hole broke the first rays of light from the ocean of light at the origin of our universe.
Time entered our cosmos accompanying the light ray which was quickly spun into atoms of matter.

As that first ray of light broke free from the ocean of light, it attempted to go from zero to light speed, but it never succeeded. That ray was quickly attracted, slowed, bent, and methodically spun into atoms of matter inside the belly of the naked black hole. At that moment, time began as the countdown of the duration during which that ray of light would spin trapped inside of an atom of matter.

As the naked black hole ate from the static ocean of light, it spun light into matter around itself. That spun light moved slower and thus lived with a duration measure known as time.

With the eventual growth to more naked black holes, more chunks of light were broken free from the ocean, and more light was spun into matter by the naked black holes. Space formed as an empty void where light had vacated the ocean. And, naked black holes devoured any freed light which was attempting to travel through newly created space.

When and where time exists in our cosmos

Over a trillion years ago, Trillion Theory states that time within our universe did not exist as yet. The ocean of light held light rays capable of carrying the time feature when released free from the ocean of light.

Once freed from the ocean of light by a spinning naked black hole, light brought time along with itself in its tool box to become active when that light would be spun into atoms of matter. (Einstein calculated that time ebbs and stops at the full speed of light at 186,282 miles per second). But, Trillion Theory states that time begins to tick whenever light drops down below the full speed of light.

Time doesn't exist in (1) the static ocean of light. And, time doesn't exist in (2) empty space or when a light rays travels at the full speed of light through empty space. However, time does exist in (3) where a black hole has broken light rays away from the ocean of light and spun those rays into matter around itself. Time is imported by spun light into the area of the sphere (3).

For whom was time created.

Here we need to think as inventors, and how the feature of time would have been designed to benefit more than one group inside of our physical universe. Time was ingeniously designed to function differently at various speeds of light.

For any planetary setting which might be habitable, the slowing of light inside of matter involved time which was designed to tick away at a normal pace so that the planet's occupants could experience the feature known as time. In this setting, growth and aging were fashionable.

But for explorers or caretakers for whom traveling our cosmos was a necessity, inventiveness had to overcome the challenge of the absolute vastness of space. Thus, time and aging could be nicely tucked away at the full speed of light.

Time on Earth.

On Earth, we don't fully appreciate the sophistication of this concept feature called time. For, time is much more than simply a clock ticking on the wall. Time is the measurement of how long light stays inside of matter in a spun state (below full light speed) as it counts down on the duration of its captivity. The speed of light slows when bent by gravity, then further slows when passing through substances such as water (rainbow spectrum), and further slows when confined and spinning inside an atom of matter.

Stated again in Trillion Theory, time is defined as the measure of the duration it takes for light which is spinning as an atom of matter to loosen and unspin to escape back to free straight line light. Therefore, time exists whenever light slows below the full speed of light when encountering the gravitational pull from a black hole or a full sphere.

Times begins in a particular area as it is released from light's toolbox when light is attracted, slowed, and spun into matter by a naked black hole. Since all fully spun spheres have a black hole at their core, they all experience and live in a time zone. Time also begins when any full speed light ray slows as it encounters a gravity pull and is bent inwards towards a sphere which has a black hole at its core.

Light attracted by a black hole is spun into an atom of matter. The various traits of light make up the various parts of the atom. The calculated duration which that strand of light stays trapped as a spinning atom is known as time.

Where time does exists and doesn't exist.

Does: Trillion Theory states that time is specific to areas where light is slowed below its full speed. Time exists where light has been spun into matter inside moons, planets, stars, and galaxies. So, time exists where atoms exist and where strong gravity slows, bends or spins light below top speed.

Doesn't: Time doesn't exist in areas where light travels at its full speed. Trillion Theory states that time does not exist in blank empty space, but light can be import time there.

Here is an example situation of how to view this: Suppose that we were inside a starship traveling at the speed of light through empty space. Even though we are matter, we are traveling at the full speed of light so time stands still for us. Suppose we passed another starship traveling at a tenth of light speed. Persons on that starship would experience time as they are moving below the speed of light and import time. The same area of space, the slower incident imports time; the full speed of light incident - doesn't.

We thereby deduce that space is not the carrier of time. The true carrier of time is a light – in its toolbox. Slow a light ray and time begins to appear in that area of space. Spin light into matter inside of a black hole and time is imported for eons of time into that particular area.

Therein, time doesn't exist in absolute unoccupied space which is empty of all matter and gravity. For time to exist in empty space, it must be imported into that area of space. Even the entry of a ray of straight line light traveling at the speed of light does not yet import time into that space. It imports only the possibility of time. Take a light ray traveling through space at the full speed of light, time stands still aboard the light ray, even though the light ray packs away and hides time in its tool chest. For time to start, that full speed light ray needs to drop below the full speed of light. That drop in speed can occur when the full speed light ray encounters a gravitational station.

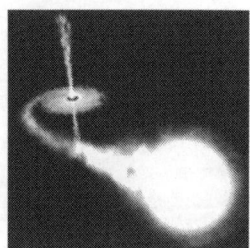

On a sun, time exists as light spins inside of helium atoms. When light escapes the sun and travels through space, time slows as the ray approaches the full speed of light, and time stops fully at the full speed of light when time hides away back inside light's tool box. However, when the ray approaches a black hole (or sphere) and slows, time is released from light's toolbox.

Time inside of a naked black hole.

We can say no, and then yes. Time didn't exist inside of the first naked black holes a trillion years ago. And, any new naked black hole tossed into the static ocean of light was barren of time. Totally naked black holes are not carriers of time, light is. Thus, light imports time into a black hole.

Therein, time is imported into the area inside and also surrounding a black hole once light is slowed, bent and spun into atoms of matter by the black hole. Time is just a feature released from light's tool box when light slows and also when it spins into matter.

Invention of specialty item known as time.

It appears that time is a specialty item created solely for this universe. Time is carried by light and sprung into action from light's tool box and imported into an area time zone when that light's speed is slowed and also spun into matter. On the other end, when spun matter unravels and escapes back to straight line light, time converts back to hide inside light's tool box. Time can hide away while light travels at full speed though totally void space (a no-time zone).

Would time exist outside of our physical universe?

Tough query. It is difficult to say one way or the other with any certainty. Trillion Theory states that time is simply an extraordinary feature invented and created specifically for our universe and its spheres. Therein, time would likely not be ticking away outside of this universe. Why would the inventors want to age if they didn't have to? Therein, it's quite possible that our physical universe was created as a destination oasis for any desiring visitant to experience the peculiarity known as time. This peculiarity of time brings with it the notion or feeling that everything we experience has a beginning and end, time makes that so.

If outside this universe is in a no-time zone or realm, then the inference of a beginning or end might not be needed. Without time, there may be simply a continuance, with no beginning or end; or some manner of moving forwards and backwards through events.

An easy clue.

To assist humans to be able to grasp the concept of time, a clue was placed for us right inside of an old man-made timepiece. It is known as the winding coil. When a winding coil clock is totally unwound and lying dormant, it doesn't measure time. However, when the coil of the clock is wound tight, time is then measured while the coil slowly loosens.

In the clock, the key winding apparatus acts like an axis and spins the coil tight around itself. Time is measured by the duration taken for the coil of the clock to unwind. Then, a human hand must come along and wind the apparatus.

Now, compare an atom. With atoms of matter, time is measured by the duration that light slows and stays spun inside of matter. Atoms conveniently have a built-in rewind apparatus (mechanism); a helix pump tightens the atom's coil whenever that it loosens. For eons the atom spins.

Time is a measure of the duration during which light's speed slows when spun into matter inside of an atom.

Unlocking more of light's incredible properties.

But, remember that light always eventually escapes from the atom and its time prison, even if it takes billions of years. Light will escape to return to travel though space at the full light speed where time stops aboard the ray of light.

In this chapter, Trillion Theory has once again shown new properties attributable to light. Each year light plays an ever increasing role in Earthly inventions. Will we further discover that light can be pushed past the full speed of light? If time slows and then stops at the full speed of light, would it move backwards at faster than the full speed of light?

CHAPTER 17
THEORY OF
EVERYTHING
(ToE)

What is a theory of everything? (ToE).

Theory of Everything (ToE) is the development of a theory which encompasses the whole gauntlet of what exists in our physical universe. This means bundling together a theory that handles everything right from micro (such as the tiny structure and origin of atoms), to the medium sized (solar systems design and their origin), and the macro (galaxy design and origin). A ToE unites all these, bundles them together and connects the dots. Such a theory might explain our entire physical universe with everything connected.

Such a daunting task demands an extremely consolidated explanation. Can the Trillion Theory of this book make such an attempt? Yes, it feels in can.

Such an attempt at a ToE has been made throughout this entire book. New theories proposed here have attempted to uniquely tie together the full scope of our physical universe of atoms, matter, light, black holes, moons, planets, suns, galaxies, space and time.

For, included in any ToE is what we can see and what we cannot see. This is on both the micro and macro levels. So far, scientists haven't been able to see everything inside a tiny atom nor inside a naked black hole or a supermassive at the center of a galaxy. Nor have they seen deep inside of a sphere to see the black hole which is at the core of every sun, planet or moon. Trillion Theory states they are there.

Trillion Theory states that the same laws and principles of recycling are seen throughout our physical cosmos. There is a direct relationship between spun matter in all the parts of our universe and on every scale of size right from the atom on the micro scale to spiral galaxies on the macro end.

Trillion Theory has done this on a simplistic model basis while offering a new paradigm way of viewing our universe. Although Trillion Theory offers very new different out-of-box thinking, its explanations are simple and easily understood. No near-impossible formulas have been used.

Trillion Theory in this book displays evidence of just how scientifically clever the design of our universe was in using only one basic materials (light) with a vast array of variety within a recycling method to perpetuate our universe.

Trillion Theory shows similarities across our cosmos:
Similarities in type of spin:
▪ spin and pumping action are in an atom which can spin for billions of years with electrons orbiting a nucleus.
▪ spin and pumping action within naked black holes creates the force to spin light into matter.
▪ spin and pumping action are in every black hole which is at the center of a moon, planet and star in our universe.
▪ spin and pumping action are in supermassive black holes at the center of galaxies. Their pumping emits plumes from the poles of the black hole forcing light back out into space.
Similarities in formations of linear shape:
▪ linear is a formation shape seen right across our cosmos.
▪ solar systems are shaped like a flat disc with a central sun.
▪ spiral galaxies are shaped like a flat disc or a pancake with a bulge (supermassive black hole) at the hub.
Note: This flat shape of a solar system or spiral galaxy is the result of the gravitational station at the hub sending out its gravity along a linear plane extending out from its equator.

Similarities in the 'selection by choice' of the direction of axial spin occurs right across our entire cosmos:

To begin, some questions:

What direction do black holes spin on their axis?

What direction do moons/planets/suns spin on their axis?

What direction do solar systems revolve?

What direction do spiral galaxies revolve?

What direction does our entire physical cosmos revolve?

Basically, are we a right or left handed universe? Is the direction of spin in our cosmos always counterclockwise or always clockwise?

The answer is that this direction of spin is determined by choice across our entire cosmos. Namely, direction of spin is similarly the *choice* of the particular black hole at the center of any sphere, at the center of any solar system's sun, and at the hub of any spiral galaxy. This choice made by any particular black hole could be a clockwise direction or it could be counterclockwise. This directional choice is made at the birth moment of the black hole, or impacted upon a black hole after it has survived a Supernova, split into two black holes, spinning away from one another. Once this spin direction happens, that black hole will maintain that same spin direction for its lifetime, however long that lifetime is within the cycle it is presently living through.

Note: Astronomers have discovered that spiral galaxies in the northern and southern hemispheres of our skies spin either counterclockwise or counterclockwise. About 50-50.

So, we see right and left-handed spin across our universe of galaxies. Obviously, the supermassive black hole at the center bulge of a galaxy sets the preference for the galaxy according to which direction it revolves.

This rotational spin is a commonality across our entire cosmos, prevalent with moons, planets, stars, solar systems, galaxies, and clusters of galaxies. It has yet to be determined which direction our entire linear universe spins.

This new innovative Trillion Theory model shows just how this commonality of rotation occurred right from the very beginnings of our universe to the present, all-across our entire universe. Rotating black holes, spinning light into matter, created the spinning feature found in our universe.

Therein, the commonality across our cosmos is that black holes have produced the multitude of spheres seen in our universe. In commonality, every one of these spheres has a rotating black hole at its core running the show.

The direction of spin can be counterclockwise or it can be clockwise for different spheres and galaxies as dictated to them by their central black hole. In that way spinning black hole spheres are independent from other nearby spheres or galaxies as to direction of their axial spin.

Trillion Theory shows similarities across our cosmos.

As a Theory of Everything, Trillion Years Universe Theory proposes that our universe known as RECYLIUN (Recycling Light Universe) has been recycling itself for a trillion years, the equivalent of 67 of the fifteen billion year cycles of our universe. The material of this recycle has been light with all of its incredible properties and the instrumental engine to spin that light into matter has been black holes.

Light and black holes are in every quadrant and zone of our universe. Light and black holes are absolute common denominators. Movement provided by light and black holes is in each and every micro and macro part of our universe.

Trillion Theory discovers a new form of life.

'Trillion Theory discovers brand new life form throughout our universe. To go where no man has gone before.'

We always thought the next form of life discovered would be aliens coming to our planet by starship or that our

SETI (Search for Extraterrestrial Intelligence) would come back to us as radio waves from some distant planet. Instead, a form resides at the cores of spheres; namely black holes.

Trillion Theory shows how black holes are 'alive' - just a different type of alive than we have ever seen before. In Trillion Theory, black holes have many life-type capabilities: they can spin prolifically; they can create a gravitational pull around themselves; they can exist at the core of a moon, planet or star for billions of years; they can hold matter around their bodies; they can hold other spheres in orbit; they can replicate; they subdivide their compartments; they grow larger; they withstand Supernovae and Obliteration never being destroyed, surviving forever; and they can recycle light and matter of our universe.

So, a black hole eats, shows greed, changes its eaten material into complex atoms of matter, holds onto that matter for billions of years, grows in size, grows in tissue, reproduces, adapts to its situation, changes its strategy methods; and has lofty goals such as becoming the galactic center of a spiral galaxy. Seems very alive.

As the machine mechanisms of our universe, black holes have been masterfully constructed, unparalleled for sure. One of the monumental tasks of astronomers in the future will be to look inside and examine the working parts of a black hole (Trillion Theory has only made its best guess).

How extensive is a Theory of Everything?

Trillion Theory really provides an individual with a new realization of our physical universe. Each time we learn more about our universe we are amazed at just how complex it truly is. Yet, we are slowly puzzling that complexity into a mosaic which makes sense and helps us to fully understand.

For instance, a person gains a new perspective of how a sun amazingly got to be a sun or a planet/moon formed.

Isn't that the goal of a Theory of Everything, to provide us with a new realization paradigm?

To my thinking, a Theory of Everything finds success if it can tie our universe together or at least show a very strong relativity between all the complex working parts.

Most spin, pumping action, and the gravitational forces within our universe go totally undetected by us. Stand and feel all the forces affecting your body at one time and you likely don't feel many of them. Yet, gravity is holding you to the surface of this planet. This planet is rotating on its axis so there is a movement of the air and atmosphere. This planet also travels in orbit around the sun of this solar system. That sun travels, taking us with it, within a spiral arm of our galaxy. Then too, our galaxy also revolves while it travels through space passing other galaxies.

Trillion Theory shows magnificence - a GRAND DESIGN.

There are inadequate superlatives to ever describe the magnificence of our universe. If it was a contest, I'm positive that our universe would win in a runaway for its simplistic yet uniquely scientific grand design.

Suppose that you were given the challenge of creating a universe and the rules stated, *'only one basic material and only one type of engine allowed.'* Such stringent rules would make the grand design for a universe extremely difficult. Of course, you would attempt to design a material and an engine as scientifically complex and versatile as possible. It would indeed be a venture to see the actual invention of those two wondrous items.

In actuality, light was designed with incredible science behind its properties, and then mass produced, as the mass material for our universe. Light, wondrous as a material and possessing so many unique properties that proper use of it affords a multitudinous variety of outcomes and products.

Light, this material which could be spun into over one hundred basic elements; able to withstand physical and chemical changes; able to exist in atoms of gas, liquid or solid and often move back and forth from one medium to the next; able to couple up as molecules to further the variety. Light, capable of forming the body of matter of all moons, planets, stars, and galaxies. Totally an indestructible material, light capable of recycling back and forth over a trillion year span, from light to matter – matter to light.

Black Holes were designed-constructed as the engines of our universe. Black holes are as unique as light, in their own way. 'If light can recycle over and over again into matter and always recycle back to light, then black holes are the engines which provide the power to make possible that recycling process.' Black holes are such incredible engines of our universe. They are engines which spun and still continue to spin the moons, planets, stars, and galaxies of our universe. And, also recycle those spheres from one era to the next. They are gracefully powerful super fast engines, capable of capturing light and spinning it into spheres and galaxies.

Trillion Theory in this book displays evidence of just how clever the grand design of our universe was in inventing and using only two basic items for that design. Both light and black holes are most worthy of being called super-duper scientific feats of the grand design of or universe.

Perhaps, the most amazing aspect of our universe is the implementation of similarity from the micro to the macro; from the simplest and the smallest atoms, to the most complex and most gigantic galaxies, similar principles were neatly applied in the sculpting. As an example, take the over 100 elements on our Planet Earth. They're all created on a similar theme: electrons orbiting a nucleus and yet lengths, thickness, tightness, density, and amounts of light used vary with each different element making for great variety. And, those elements have different states as a gas, liquid or sold, and ways of reacting chemically and combining with other elements to make molecules.

Inside the absolute cleverness we can see the strategic techniques. For once light and black holes were put to their assigned tasks to cosmos build, their interaction required no further outside manipulation. Light and black holes grew our universe larger and larger by recycling moons, planets, stars, solar systems, and stars within galaxies for one trillion years.

What to make of the inference of a GRAND DESIGN.

Humans are at an exciting point in existence. We have learned how to think more for ourselves with each new generation. However, science and church are still distant in cumulative attempts to figure out our universe.

But, before getting labeled, please allow me to state my position: I am an advocate of a scientific perspective and scientific proof. Yet, I do believe in an individual having something that could be called a soul. Read my 2013 novel *The Trillionist* for my thoughts on that subject.

Yet, I attribute everything to science. Even a person's soul, should it exist, is a grand scientific achievement by whoever or whatever did the inventing, design, and building.

Please permit me to elaborate. It is my personal believe that our universe is complex beyond belief. Why shouldn't it be? It has every right. Yet, my theories show the simplicity in materials and construction methods deployed in building. My theories demonstrate an ultra-strategy and cleverness behind our universe's design – namely complexity hidden behind simplicity. Yet, do such thoughts of a grand design make me into a creationist? No.

Do we need to include who and why in a ToE?

As an end goal, that is man's final determination – why are we in this universe? What is life about? This book and Trillion Theory do not aspire to resolve creation. Rather, Trillion Theory deals with our physical universe. But stating how clever and strategic certain physical things are in our universe is not creationism, rather it is the responsibility of Trillion Theory to point to the grand scientific methods used.

Think about it! Every time we move to a new level of understanding of our universe, we uncover greater ultra complexity. The discovery of atoms took us from simple to ultra complex. It is my contention that the uncovering of the ultra complexity of our universe will continue to escalate. As we discover more about our physical universe, we will find more complexity at each new level. We have yet to uncover all the wonderment of light, black holes, and supermassive black holes controlling galaxies, as my theories will disclose.

Further on the complexity notion, I also contend that any thought of a divine omnipotent creator behind the wheel of our universe will also move to a much higher ultra complex rung after we can fully understand our physical universe. Whether we call it a divine omnipotent creator, or an artisan, architect, corporation, or our universe existing as a computer hologram program, all this is bound to become more ultra complex and follow the same pattern and road which we are traveling as we understand more about our cosmos. Trillion Theory shows our universe as one finely oiled machine, purposefully operating and scientifically designed.

Yet, I don't endear *creationism* which advocates that all things were created as they now exist. With that I can't agree. I believe in evolution. Trillion Theory takes evolution a step further, proposing our universe of spheres evolved and grew in size and numbers over a trillion years of history.

Creationists believe everything was created from nothing as described in the bible. Instead, this book shows how all the matter of our universe formed from huge amounts of an energy source which was and still is in ample supply.

Trillion Theory submits that a trillion years ago something was obviously behind the wheel designing our physical universe in a preplanned clever strategic recycling manner. This growth occurred on cue from universe laws or cosmic rules specifically and cleverly implanted pertaining to the formation and the growth of spheres and galaxies.

Trillion Theory demonstrates just how strategic cleverness was used to build our ultra complex magnificent universe. As a marvel beyond any comparison, our universe is a living growing entity incredibly perpetuating from one 15 billion year cycle to the next, all according to a strategic system.

The question arises: Is our universe mere happenchance, purely an accident? Or, did the hand of some creator-artisan or architect place scientific design onto its construction?

Throughout, this book resists placing a creator 'at the switch' or 'behind the wheel' and deals with the intricacies of the observable knowable physical universe.

When you read this book, your thoughts as to a creator-type-entity might change, and whether my theory influences any change in that regard is not my direct intent. From my perspective, I want to sit on the fence and keep my mind open. Our ultra complex universe may yet have a totally far different design component we have yet to discover.

CHAPTER 18
WHERE'S THE
PROOF

The Big Bang had better come up with better proof than, 'Galaxies seen through telescopes are receding away from one another, meaning there must have been a big explosion creating our universe.' That is insufficient proof of Big Bang.

Trillion Theory flatly states that current widely accepted proofs of the Big Bang are not sufficient proofs. Collected scientific evidence has been twisted in an effort to make everyone accept Big Bang as the origin of our universe.

Also, it is not sufficient for astronomers to measure the age of older stars at 13.7 billion years and conclude that's how old our universe is. Suppose someone saw the oldest trees in a forest at 200 years, should they simply conclude that the forest is 200 years old, when in reality it is millions. The difference is that our universe is the ultimate recycler, hiding the 66 past 15 billion cycles of our cosmos.

There are 3 things to accomplish here:

♦ Show that Big Bang theory is a false unproven theory.

♦ Show how Trillion Theory is a possible correct theory and how it tackles the claims made by other theories.

♦ Provide possible proofs for Trillion Theory, and suggest ways that astronomers and astrophysicists can assist.

Say 'no' to Big Bang hoax. Getting rid of Big Bang.

Big Bang theory, not good enough for acceptance today.

If Big Bang is an incorrect kaput theory, it belongs right alongside the ancient false flat Earth belief (proven wrong as Earth is roundish), and another theory which lasted till 1600, namely Earth as the center of our universe (proven wrong as our sun centers our solar system, one of millions of solar systems inside of our Milky Way Galaxy).

However, Big Bang Cosmology is still promoted today as an established fact and taught it in schools and universities. That theory is strongly engrained in millions of people. So it must be right? But, over the past 60 years, discoveries have often been twisted so that they'd fit neatly into Bang.

But, the Bang has everything wrong as to the origin, age, and the formation of the spheres of our universe. It claims that our universe began at a source point which exploded, expanding across space some 13.7 billion years ago. Since Big Bang had no way to show how spheres, solar systems, and galaxies formed, Nebular Theory was recruited to state how gas clouds supposedly whirled, eddying into spheres.

But, Trillion Theory adamantly says 'no' to Big Bang. Yet, it is necessary here to examine Bang's claims since nearly every reader has been taught to accept the Bang as law. So, before there can be a new sheriff of universe theory, either Bang has to be proven wrong, or a new theory proven right.

Paradigm shift to a revolutionary new cosmology model only occurs when one new theory replaces old thought and the new theory brings some paramount scientific proof.

Declaration: Big Bang's simplistic depiction of the origin of our cosmos is a false and outdated incorrect theory. It is based too heavily on happenchance dreadfully failing any test to properly explain many complexities of our universe. Big Bang simply leaves such subjects to one's imagination. Note: Renowned pioneer Radio Astronomer Grote Reber had always been skeptical, stating that Big Bang was bunk.

Why Big Bang Theory refuses to go away?
Many huge gaps and weaknesses in Big Bang theory.

There is difficulty in challenging an entrenched paradigm such as a Big Bang. For the past 60 years, astronomers and astrophysicists have been working their butts off trying to find some monumental proof of Big Bang. They will argue they have proofs: redshift galaxy readings indicating galaxies ebbing away; microwave background radiation indicating that our cosmos began very hot leaving behind a microwave glow; and gravitational waves indicating past expansion. But, Trillion Theory refutes these Big Bang claims as inadequate.

So, in today's modernistic world, one would think that it should not be that difficult to know when an old theory has been *blowing smoke* for too long. It should be relatively easy to replace an old incorrect theory such as the Big Bang?

Problems had already begun to surface for Big Bang

A few years ago, astronomers made the discovery of a star which they calculated as 18 billion years old. In 2013, this paradox continued unsolved with the discovery of the impossible Methuselah Star, at 16 billion years ancient. If the Big Bang origin of our universe occurred 13.7 billion years ago, how could two stars trump the age calculation of Big Bang? Definitely a paradox and a fly in Big Bang's ointment.

Big Bang relies on false premises, Trillion Theory says:

• **False to Big Bang Theory** which depicts our universe age at 13.7 billion years. This is an inaccuracy resulting from only estimating the age of presently visible stars in our sky.

• **False to Big Bang theory** depicting the origin of our universe from a central explosion of matter outwards.

• **False to a Big Bang** conclusion that galaxies receding from each other proved an explosive origin to our cosmos.

• **False to Inflationary Theory** attempting to tag onto Bang stating that our universe expanded exponentially fast.

• **False to Nebular Theory's** explanation of how planets, moons, and stars formed. They didn't form as a result of a Big Bang or from Nebular swirling gas clouds.

• **False to Nebular's** hypothesis that size mattered in the cooling process of spheres as larger-in-size stars fended off the coldness in space for billions of years, while the exteriors of smaller planets/moons cooled forming hard surfaces.

Attacking Big Bang's three main supposed proofs:

Attack Big Bang's supposed first proof: redshift galaxy readings indicating that galaxies are receding away from each other is interpreted as Big Bang's proof of an explosion origin. Bang Theory first came about as a theory because of this redshift discovery. Simplest conclusion at that time was that a central explosion sent our universe's matter outwards.

But, Trillion Theory contends that our cosmos is being pulled outwards by an ocean of light which exists on the perimeter surrounding our universe. Galaxies are continually attracted towards this perimeter, thereby always expanding space's outer boundaries in an ever expanding universe.

Big Bang's supposed 2nd proof: microwave background radiation supposedly indicates that our cosmos began very hot leaving behind a microwave glow. Trillion Theory offers up other reasons for this radiation. Spheres have recycled over and over, and each time light spins or escapes, it emits radiation. Thus, microwave radiation should be everywhere. Trillion Theory states that Supernovae occurring over the past trillion years left behind microwave radiation in the aftermath graveyards of billions of obliterated solar systems.

Big Bang's supposed 3rd proof: gravitational waves might indicate past expansion right after a supposed Big Bang. Recent discovery shows ripples in time-space which the top astrophysicists think are the very first tremors (aftershocks) 380,000 years after the Big Bang.

Trillion Theory refutes this as non-proof of Big Bang and adds that the ripples should be there. They are leftovers from a trillion year history which witnessed small galaxies totally exploding when their massive sun went Supernova and exploded its entire galaxy. These were massive recycling explosions, leaving behind time-ripples in space.

There never was a Big Bang start to our universe!

Trillion Theory declares that Big Bang never happened. Trillion Theory demonstrates how solar systems and galaxies are structured to the nth degree. Organizational laws repeat themselves in all solar systems during a trillion year history.

As for the Big Bang, Trillion Theory contends that the Big Bang is a fallacy – built upon pillars of sand. When future generations realize the true origins of our universe, they'll look back to today's Big Bangers and laugh hard, just like we view those of ignorance who once believed in a flat Earth.

Today, lack of a broader understanding of our universe is holding humans back in a multitude of ways. Big Bang has only sufficed by being a convenient explanation. But, there exists an absolute need for the next generation of scientific discoveries to uncover more universe secrets.

Trillion Theory attacks Nebular Theory too.

Nebular attempts to assist Bang in addressing how solar systems formed. Supposedly Bang's gaseous nebular clouds coalesced. Stephen Hawking states that stardust of swirling nebula clouds saw a central gravitational force assemble all the gas particles into spheres in our solar system.

Trillion Theory says Nebular Theory is as ridiculous as the Bang. For instance, both Bang and Nebular can't explain the radically varying axial tilts of the planets in our solar system.

Also, after the supposed Bang occurred, how come stars don't contract in the extreme coldness of space, but rather expand? Ten billion years is an extremely long cold time.

How come volcanic flows on planets still fire outwards, when a planet's core should actually be freezing in space's cold depths? How come spheres don't simply freeze solid? Bang supporters say, compression of the surface started the fire in the pit of planet Earth. Trillion Theory won't buy that.

The real question should be: what internal force is really causing the interior of that planet to heat up and get hotter?

A planet sitting in the dark extreme coldness of space. What really started the fire deep in the bowels of its core? What expanding force is sending pressurized lava to the surface?

Nebular Theory **incorrectly** premises our solar system as having begun some 4.5 billion years ago as a large irregular cloud composed of gas and dust. But, Nebular does nothing to explain where elements found on Earth came from.

Nebular theory **incorrectly** proposes that a gravitational collapse pulled a gas cloud close together to form our solar system, forming collection spherical clumps. Nebular has no way of accounting for where gravity actually came from.

Nebular **incorrectly** premises that the largest clump coalesced to form our sun. Hypothesizing that where the most mass gathered, it became hotter to form a sun.

Trillion Theory attacks Inflationary Theory.

Inflation by Alan Guth tried to answer a classic Big Bang conundrum: why does the cosmos appear flat? Big Bang's explosion should have been isotropic (in all directions). Trillion Theory purports our universe as linear, not roundish. But, offers a totally different reason for this linear than the supposed inflation explosion put forth by Guth.

Trillion Theory disagrees with Stephen Hawking.

There have been admitted mistakes by those renowned.

Even Stephen Hawking, the admired and renowned English theoretical physicist and cosmologist, in April 2013 admitted to a rather large blunder. Till then, he'd thought that light swallowed up by a black hole was lost forever. Hawking recanted his stance admitting that his wondrous discovery '*Hawking Radiation*' does in fact escape from a black hole. This agrees with Trillion Theory showing light escaping our sun after billions of years of entrapment, finally eluding the clutches of a black hole at the sun's core.

Hawking further states 'radiation is emitted from the rim of a black hole, where all the matter that is falling into the abyss is stretched out like spaghetti. Even light does not escape from a black hole.' He is partially right, then wrong. His stretching light out like spaghetti is in Trillion Theory.

Show Trillion Theory (TT) as the correct theory. All other theories incorrectly show a 13.7 billion year universe.

TT declares that incorrect 13.7 billion year age estimate by other theories is a result of them focusing solely on the age of the older stars in our present sky. TT says dig deeper, and find that the stars have recycled many times over giving us the true age of our universe at a trillion years since origin.

Why proof of Trillion Theory (TT) is vital?

Because Trillion Theory covers so many universe aspects, there will be numerous ways to prove Trillion Theory.

Trillion Theory states that black holes over the past trillion years of universe history are the builders, the organizers, and the operators of the spheres, solar systems and galaxies of our cosmos. Stated more directly, 'Trillion Theory shows that black holes, as expert recyclers over the past trillion years of universe history, are the builders of spheres, the operators of solar systems, and the organizers of galaxies in our cosmos.'

Proof of Trillion Theory is vital for man's next steps on Planet Earth and also out into the realm of spheres. For now,

TT is just mere conjecture - great fodder for all discussions.

So give Trillion Theory a chance. Trillion Theory deploys a more advanced paradigm model depicting a totally different way our universe began, grew and recycled over the past trillion years. And hopefully new scientific methods will provide proofs. For at the outset, a theory is a proposed explanation done as a theoretical hypothesis stating a definitive belief about a phenomenon in our universe. When a theory is proven, it becomes accepted norm, the new PARADIGM.

Proofs for radical new Trillion Theory (TT).

Always, science is called upon as the means to prove or disprove a conjecture theory. It is one thing to say it is so; it is quite another project to prove or disprove a new theory.

The biggest obstacle to proving theory is the short time of man's existence compared to a Trillion year history of our universe. Man's time goes by in a blink of an eye, while our cosmos recycles at a snail's pace. As actor George Clooney, of the movie *Gravity,* said about man's existence compared to the cosmos: "I'm very aware of just how brief life is."

For instance, to catch the action at the start and end of one solar system we'd need to run a video for 15 billion years. So instead, we are relegated to piecing together the snail-like action from hundreds of different solar systems in various stages of their cycles. The good news is we humans are very resourceful. As Carl Sagan remarked, 'Extraordinary claims require extraordinary evidence.'

That was the case in November 2013 when American Press announced, 'Earth-like planets fill our galaxy.' A study, published in the National Academy of Science, found our Milky Way is teeming with billions of planets circling stars similar to our sun. Trillion Theory had already clearly stated that solar systems are the norm throughout the cosmos, and a star without a solar system is considered an abnormality.

Possible proofs to authenticate new Trillion Theory (TT).
Proof Method 1 for Trillion Theory (TT).
(Black holes in Supernova-Obliteration graveyard).

Difficult, but doable. Powerful telescopes may someday peer deeper into landscapes after a Supernova. TT contends there will be many naked black holes (remnants from the destroyed spheres of the old solar system) occupying the area and filling up with light to begin the next cycle.

Powerful space telescopes have difficulty seeing into the brightness of a Supernova and the Obliteration of a solar system because of the huge flash of light. Just as hard to see is the dark blackish holes left behind right after Supernova and Obliteration. TT says that survival black holes dominate the graveyard after the death of a solar system.

If many black holes are found in the aftermath death of a solar system, then that proves Trillion Theory. TT details how the black holes have reproduced and are now preparing to rebuild that solar system into the next 15 billion year cycle. However, that graveyard might also be very bright instead of pitch black, as the black holes in that area will be pulling in streams of light and spinning them into matter around their bodies. The black holes may be hidden behind veils of light.

To find this proof, an astronomer would really have to be Johnny-on-the-spot, for blinking or being late to the party would mean missing those naked black holes which would have gotten the jump and filled up again to appear as new moons and planets. Too late and each black hole would be hidden away once again within a sphere's core.

Proof Method 2 for Trillion Theory (TT).
(Axial spin direction varies with solar system spheres).

Nebular Theory incorrectly states that every sun, planet and moon formed by a certain nebula should have the same direction to its axial spin. But, they don't.

If the nebular cloud supposedly spun coalescing to form our sun and planets, then that spin should have given all the planets the same direction of spin on their axis, but it didn't. Nebular can't explain why certain planets and moons rotate clockwise while others rotate counterclockwise on their axis. For example, Earth spins counterclockwise on its axis, while planet Venus rotates clockwise.

Now, let's investigate bizarre Venus. It is most likely a mirror sister of our planet Earth. When, the old solar system which used to occupy this area was obliterated by its old sun going Supernova, one of the average sized black holes of a planet survived and split its black hole into to two near twins: Earth (7,926 miles in diameter); Venus (7,521). Venus and Earth are a little different in size simply because their black holes ate and spun light into matter at different rates. Fierce splitting spun their two black holes away from each other, applying two directly opposite rotational directions: planet Earth spun counterclockwise, while its sister planet Venus rotates almost perfectly clockwise.

This discrepancy as to differences of direction of axial spin is well explained in Trillion Theory. After a Supernova within any solar system, the splitting which occurs inside the surviving black holes applies directly opposite rotational spin directions to the two new black holes.

Proof Method 3 for Trillion Theory (TT).
(Axial tilts of solar system planets differ).

Nebular Theory incorrectly states that every planet which was formed by a certain nebula should have the same axial tilt; within a solar system they all should have the same axial tilt when riding in orbit. But, they don't.

Those varying axial tilts of the planets in our solar system should not be there if there had been a Big Bang followed by Nebular. If the nebular cloud supposedly spun coalescing to form our sun and planets, then that spin should have given all the planets similar axial tilts, but it didn't. Nebular fails to explain why the axial tilts of some planets differ.

Astronomers have shown that in our solar system the tilts of planets differ, and nebular theory has no way to explain. Trillion Theory shows our solar system's planets as operating under their own tilt. Every sphere in our solar system has an axial tilt unique to itself. This is proof of Trillion Theory.

Trillion Theory shows how each black hole (which formed a planet) spun at different axial tilts right after the old solar system which used to be on this spot was destroyed by the Supernova-Obliteration death of the last cycle. Varying tilts for each new black hole occur, as each black hole will search for its best battling strategic angle for attracting light in order to build a sphere around itself.

So now, each planet sits in orbit in an axial tilt which was decided by its black hole back when it was trying to find its optimal angle to allow it to best fight for light. This axial tilt has remained with the sphere even after it found its orbit around our sun. Example: Earth's axial tilt is 23 degrees.

Now over to weirdo Uranus, which spins on its side because according to my Trillion Theory all black holes can independently determine their own axial tilt when they are battling competitor black holes for light to spin into matter. Uranus is direct **disproof** of Nebular Theory and Big Bang. Uranus axial tilt is direct **proof of Trillion Theory.**

Proof Method 4 for Trillion Theory (TT).
(Find the black hole at the center of a sphere).

If we were to dig to the core of any moon, planet or sun, Trillion Theory proclaims we would encounter a spinning black hole. This would be direct proof of Trillion Theory.

However, the black hole which we would find at the core of our moon for example, would appear much different than any distant naked black hole astronomers have perceived through their telescopes. For, the full black hole at the core of our moon has slowed its spin. Eons ago it became sluggish when it filled its core and all its compartments with heavy matter. It'd be difficult to differentiate the moon's black hole from the rest of its outer body; for the moment, in this part of the moon's cycle, black hole and body are one. The black hole gives the moon its direction of spin on its axis, and holds matter on the moon's surface. And the black hole holds the early fire which will someday make its way to the moon's surface.

Yet, it would be an ultimate rush to one day travel down into the inner depths of a sphere, to dig deep into its black hole. In Earth's case, we'd need to somehow overcome the intense heat where the oldest atoms have unraveled, created a fire, and already reached Earth's surface via volcanoes.

Proof Method 5 for Trillion Theory (TT).
(Spin light into matter in a physics laboratory).

The 5th proof of TT may be possible right here on very our own planet, and prove TT beyond a shadow of a doubt.

In the medieval ages, alchemists the pseudo-scientists of that bygone era, tried to make gold from various materials. They failed bitterly. Yet looking back, it was a valiant effort. They attempted a feat never accomplished, to create matter.

In my futuristic sci-fi novel *The Trillionist,* the young lad in the novel invented a machine called a Quantronix which artificially did the task normally assigned to black holes. Quantronix was a very large fat machine with a long giraffe-type neck which stuck out the top of a domed building. That Quantronix attracted light from the sun, bent it, and then spun the light at great velocity into atoms of the purest rare elements. Many different types of elements were formed, depending upon the length of the strands of light cut; and the tightness of the spin of the light into atoms of matter. The young lad succeeded at becoming our universe's first inhabitant to create matter from light. The problem was - would this discovery be used for good or evil.

Perhaps in the future, humans will invent a Q-machine, to become artisans of elemental matter of this universe. Such a discovery would open up numerous hidden secrets about our universe. Such achievement would prove Trillion Theory, furnishing proof of how black holes formed our cosmos.

How modern day astronomers and astrophysicists can assist with the finding of proofs for Trillion Theory.

In the future, astronomers and astrophysicists will have a plethora of opportunities to assist in proving Trillion Theory.

Astronomers are learning more about the order brought to a galaxy by the supermassive black hole at its hub. But, they struggle to find rhyme or reason to smaller black holes. They ask, 'why the two types' without being able to answer.

It's incorrectly thought by those peering through huge telescopes that black holes are only formed as the remains of a sun gone Supernova, as it explodes and then implodes. They contend that this inward implosion is so powerful that a star collapses inwards forming a black hole. This provides a false reason for a black hole after Supernova.

It wasn't ever rationalized till Trillion Theory that the black hole initially had formed the star around itself. Thus, when the star went Supernova and exploded away all its matter that black hole survived and became nakedly visible.

Trillion Theory shows that there to be more than one type of black hole in the cosmos. Yet, remember that black holes are difficult to see in the pitch blackness of space, so many are never spotted doing their work. But, astronomers are eagerly studying black holes, this will help Trillion Theory.

CHAPTER 19
SUMMARY
AND
CONCLUSION

Summarize the Goals of Trillion Theory:
Main Goal (Mission) of Trillion Theory – a new theory.
The main mission of Trillion Theory is to propose a plausible new theory depicting how our cosmos originated one trillion years ago, then grew to its present prodigious size. And focus on the major role played by black holes in this growth. This mission seeks to radically jump-start a game changing view of our universe into a new Trillion Theory paradigm.
Second Goal of Trillion Theory – Proof.
Trillion Theory offers in this book 5 possible ways to PROVE its theories using future science. Thereafter, to have brand new Trillion Theory become the NEW PARADIGM model for properly understanding and explaining our universe.
Third Goal – '*No*' to Big Bang.
Declaration: Big Bang never happened 13.7 billion years ago. It is high time to rid ourselves of the old Big Bang, expecting it to explain our universe. It is a 60 year old theory, hanging on by its fingertips - a sacred cow waiting to be tipped over. Big Bang fails all tests, offering no real true proof. Bang is far too simplistic to explain our universe. In reality, Bang Theory came about because of the discovery that our universe is expanding with galaxies receding away from each other. The simplest conclusion was that a central explosion had sent all matter of our universe outwards. But, Trillion Theory does suggest other reasons for this receding and expansion.

Short Summary of Trillion Theory in this book:

• Trillion Theory sees an ordered system – not chaos.

• Trillion Theory sees 'orderly arrangement' in our universe resulting from cosmic laws of engagement between spheres. Laws seen throughout billions of solar systems and galaxies.

• Trillion Theory asserts that our universe began small a trillion years ago and then grew to its present gigantic size.

• Trillion Theory expounds that the history of our cosmos has been one 15 billion year cycle after another; each new cycle intertwining with the last; growing grander with each cycle to our present 67th cycle hosting 73 quintillion stars.

• Our universe's history has seen 67 cycles during its trillion years, with each of these cycles spanning 15 billion years.

• During a cycle, stars age out by going Supernova thereby recycling themselves and their solar systems into new ones.

• The spheres and solar systems in our present sky are just the latest rendition to occupy our current 15 billion year segment cycle of universe history.

• Spheres and solar systems recycle about every 15 billion years, while the galaxies which they reside in can be much older because of the older supermassive black hole at the core of the spiral galaxy.

• Trillion Theory suggests that our cosmos originated from an endless ocean of static (frozen-like) light.

• Trillion Theory shows how the endless ocean of light has over history, and still is, the supplier of the light being used as the material to make the matter of our cosmos.

• Trillion Theory states that the billions of spheres, solar systems, and galaxies all originated from this ocean of light.

• Trillion Theory describes an outside force pulling the contents of our universe ever outwards towards the ocean of light now located on the outer perimeters of our cosmos.

- Light spins to form atoms (elements) of universe matter.
- Light carries in its tool box all the properties for matter.
- Light locked away as an atom of matter will spin for eons.
- Light will always want and succeed in someday escaping its atom and depart away at the speed of light.
- Naked round black holes are the power engines designed with spin speed methodology to spin light into matter.
- Trillion Theory shows how black holes spin forever, on and on, it's what they do. And on the micro scale, they also spin the atoms (elements of matter) during their sphere building.
- Naked black holes spin matter to fill their belles and to form planets, moons and suns around their spheroid shape.
- Trillion Theory shows black holes (machine-like entities) as the hard-working builders of all the spheres of our cosmos. These indestructible black holes are also the organizers and controllers of billions of solar systems and galaxies.
- Therefore, Trillion Theory advocates that a black hole exists at the center core of every planet, moon and sun (star).
- XL planets have the possibility of evolving into a sun. But, most never get the chance.
- Neither light nor a black hole can ever be destroyed. They are the two main indestructibles of our recycling universe.
- Black holes reproduce, splitting via fission from one to two when a sun goes Supernova (explodes) and obliterates the planets and moons of its solar system. The black hole at the center of each sphere survives the Supernova/Obliteration.
- Solar systems grow larger each recycle to where they can hold hundreds of planets/moons in orbit around a sun(s).
- One solar system can be broken apart during Supernova of its sun, thereby two new similar sized solar systems result. As well, the overall total number of spheres doubled.

• Size matters, as larger black holes overpower smaller black holes to gain control of solar systems as their suns.

• A galaxy becomes the hotel island oases of the cosmos. They provide the template holding many solar systems in place around a hub. Within the galaxy, solar systems recycle to thereby increase the number of spheres in that galaxy.

• As solar systems increase in number in a certain area, the most powerful supermassive black hole of that area takes over their hub forming a galaxy. That supermassive black hole can survive for hundreds of billions of years, while the suns and solar systems within its galaxy grow in numbers.

• Spin, provided by the super spin of black holes, is the main motion imparted to our universe. This spin is seen in black holes, planets, moons, suns, stars, galaxies and even atoms.

• Linear is the main configuration between the spheroids of our universe. This linear configuration is seen with the rings around planet Saturn, with the pancake shape of our solar system, with the Frisbee appearance of spiral galaxies, and with the elongation of our entire universe on a linear plane.

• Space is a void empty derivate byproduct left behind when huge amounts of light depart an area. This departed light is spun into matter around a black hole forming a sphere.

• Time is imported into any area where light is slowed and spun into matter. Time is the duration measurement during which light stays spun as atoms round a black hole.

• Trillion Theory proclaims our universe was 'placed on the clock' (imported time) when black holes first spun light into matter to form the first early spheres of our universe.

Conclusions made by Trillion Theory (TT):

• Trillion Theory suggests that the building and recycling set-up of our cosmos can be likened to a large project: The obvious assignment was to build the most complex universe while utilizing only a very few ultra-versatile indestructible materials, namely: versatile light carrying a whole myriad of stupendous properties, including the ultimate of being spun into matter; and black hole mechanisms capable of spinning light into a mosaic of elements and spheres. Then, the final step of the design project was to establish definite cosmic laws of engagement between light and black holes, and also between black holes in combat and co-operative situations.

• Trillion Theory suggests that our cosmos was a most highly planned project, utilizing absolute masterful technology and science for the design and the methods of operation.

• Trillion Theory suggests that a similar deployment of top level science was placed into the growth and recycle module for our cosmos, to such an extent that the recycling process would be able to function of its own accord onto perpetuity.

• Trillion Theory suggest that our universe is a technological marvel involving: strategic purpose; unique clever design; the invention of two indestructible materials (light and black holes); the spin feature built into black holes so they could provide spin all aspects of our cosmos; strategic simplistic appearances hiding complexity underneath (a macro sphere hiding a black hole; and a micro atom hiding spun light); cosmic universe laws built into black holes; organizational laws for solar systems and galaxies; recycling growth feature for spheres, solar systems and galaxies; and instructions built into light and black holes so they recycle to perpetuity.

• Trillion Theory suggests that the designers of our cosmos left several clues to help inhabitants decipher our universe, as we reach various levels of technology: (a rainbow shows a tiny sample of light's unique properties; lightning releases its powerful light and heat; an atomic explosion releases the power of an the atom showing what is inside; a Supernova releases the huge power of an exploding star).

If Trillion Theory is true, there are huge implications:

• The implications of a trillion year old cosmos could be dramatic for humans both on a physical and spiritual plane.

• If there have been living beings on planets during any or all of the other 66 previous 15 billion year cycles, then life may have abounded during a trillion years of history.

• If every star has a solar system, as Trillion Theory suggests, this greatly ups the possibility of life on many planets which have a Goldilocks situations of not too hot, not to cold.

• Some past civilizations may have reached incredible levels of technology, hundreds of billions of years ago.

• Civilizations which reached incredible levels of technology obviously discovered that their solar system would someday recycle, therefore they would have had to settle colonies outside of their own solar system in order to increase the odds of the perpetuation of their society.

• And for those who believe in reincarnation, Trillion Theory takes the possibilities to new heights. Read my novel *'The Trillionist'* to discover that reincarnation has a deep and rich history over the trillion years of our universe. Reincarnation might have the propensity to be a gigantic feature, as any spirit may have reincarnated thousands of times.

Query: When light comes from our sun to our planet Earth, the black hole inside of our Earth can recycle that light?

Author's answer: No. The Earth, at its present stage can only absorb or reflect light. For light to be spun into matter, the planet must be at a far earlier stage with its naked black hole devouring and spinning light into a sphere of matter around itself. The black hole living at the core of Earth has already completed that task, is plumb full, and now is in a matter-holding pattern. Thick matter on the surface of our planet blocks the spinning of any more light into matter.

Who helped me out with my new Trillion Theory:

No one. Trillion Theory is solely my doing.

Author's Disclaimer: Theories in the book were developed solely by the author Ed Lukowich. These Trillion Theories are not based upon any other person's theories which may somehow find any similar basis. Therein, no other person may claim to be the originator of any of these Trillion Theories.

I've freelanced, unencumbered to write my own universe theory, by always perusing new info about our universe and interpreting those clues in my own way. Many of my ideas came to me from my strategic thought processes as to how a universe might be best constructed. Example: Astronomers discovered black holes devouring light. I gave new meaning to black holes, their purpose in our universe, and how they operate to build and grow our cosmos. With light, I took a rainbow as a clue that light could bend – and then spin.

However, I've had many heroes of history to look up to: Copernicus, Galileo, Newton, and Einstein; and then Carl Sagan who had this wonderful quote: 'It is far better to grasp the universe as it really is than to continually persist in delusion, however satisfying and reassuring.'

Yet, it pleases me to live in a world where those smarter have gained insight to explain of our cosmos. Truth has always emerged victorious, even if taking decades or centuries. Inside me, there had always been an innate need to better understand life. Like many people, my eyes have looked to the sky, wondering what this world is about.

Today, I see our night-time sky much differently than I did as a child. Today, I see spheres made of incredibly tightly spun light-to-matter. And, then there is the endless empty space left behind as all that light departed into the spheres.

The original ocean of light, from a trillion years ago, is now gone from our sky. Instead, in our viewable sky, all light (tons and tons of light) is presently tightly packed as matter inside of the sky spheres. Space has replaced the area which the static ocean of light used to occupy. These huge areas of space are just the weightless, dark, cold void byproduct left behind when the black holes spun all the occupying light into tight tiny atoms of hard matter. The light which spun into the spheres, took with it all the weight, heat, and other properties and left space behind as a voided emptiness. Clever, as light erected space to be used in the future as a highway to travel after light finally exits and escapes from the clutches of the black hole at a sphere's core.

Mastermind physicist Albert Einstein's work helped me tons. His formula E=mc² (or mc²=E) showed unbelievable amounts of energy stored within an atom. Energy and matter had this back and forth *forever* relationship. Matter and energy recycle back-forth but they can't be destroyed.

That formula is now engrained into Trillion Theory. So when Supernovae occur, matter escapes back to light trying to depart the scene, and when a naked black hole captures and spins that light, it returns to being matter.

To make my point, imagine we blow up Jupiter. Planet Jupiter is 88,736 miles in diameter, that's a lot of mass. But, here's the key, every atom of Jupiter will simultaneously split and explode. We know of the staggering release of energy from a single atomic explosion; *m* (mass) multiplied by *c* (light speed 186,282 squared) = E (967,000,000,000,000,000). The pop that genie releases is tremendous pent-up energy.

Quick ballpark math: energy (light, heat, radiation) from Jupiter atoms exploding takes up 1 million X 88,736 miles as the diameter of Jupiter = 88 billion miles). The release of all the atoms of Jupiter takes up far more area than the 7.5 billion miles diameter our entire solar system occupies.

The main point is that a moon, planet, or star is more than just a hunk of matter. My theory states, 'What comes out is what originally went in.' The amount which comes out is jumbo. On the reverse side of the formula, massive light amounts were spun to make each cosmic sphere; each time leaving behind endless stretches of space. Unfortunate for us, so far on Earth we have figured out how to split an atom, but have yet to discover how to build an atom.

Sayings which come about from new Trillion Theory:

Trillion Theory: 'Someday we may climb high enough to peak over the fence into the Artisan's backyard.'

'Nothing worthwhile is easy, least of all Trillion Theory.'

Carl Sagan: 'No great idea ever started without first being thought of as ridiculous.'

Final author comments:

Thank you for reading this book. My ultimate hope is that you now think 'Trillion' and 'Black holes,' as I do. Or, some corridors have opened for you to new ideas. For, tweaking the theories in this book has been a gargantuan task, thwart with enormous complexity. In the near future, I will seek opportunities to take Trillion Theory through other publicity mediums, which is tantamount to achieving success.

On the flip side, critics will provide for a re-examination of ideas. No doubt, my new Trillion Theory will be bashed and trashed by many. And that's good. My philosophy is that the person who opposes often helps as much as a supporter.

'My passion outweighs my fears.' Of course, criticism is certain to come my way. 'A risk I see as well worth taking.'

My intent is to have no writer's regret. For, my Trillion Theory is either right or wrong, accepted or rejected on its own merit. In the long run, new scientific discovery is the litmus for proof. For that exact reason, we should never have placed an unproven Big Bang on such a high pedestal.

Here are some quite recent dramatic examples where man has had moments of being wrong: Western Union 1876 memo read, 'This telephone has too many shortcomings to be seriously considered as a means of communication;' Lord Kelvin 1899, 'Heavier than air flying machines, impossible;' Charles Duell 1899, U.S. Patents Office, 'Everything that can be invented has been invented;' Robert Millikan, Nobel Prize for Physics 1923, 'There is no likelihood man can ever tap the power of an atom;' Thomas Watson 1943 chairman of IBM, 'There's a world market for maybe five computers;' Ken Olson 1977, founder of Digital Equip, 'There's no reason anyone would want a computer at home.'

And more often than not, it takes time to install a new paradigm theory. It was 150 years, from the time Copernicus first whispered that the Earth was not at the center of our cosmos, to the acceptance of his new theory which placed our sun at the center of our solar system.

So again, thank you for reading. Please keep up with my present website **www.trillionist.com** where there will also find a link to my future website *The Universe Trillion Project*.

I'm delighted that you take a universe interest. Hopefully my Trillion Theory is either your new theory too or at least it has opened up some new thought corridors.

The future will definitely bring many new amazing cosmic discoveries. In 2018, NASA launches the James Webb Space Telescope as its new premiere space observatory, with two main scientific goals: to further study the formation of stars and planets and to study the birth and evolution of galaxies. Also, there is that eternal hope of discovering alien life and other Earth-like planets. And, from this author's perspective, to hopefully discover new evidence to validate, authenticate and prove my new Trillion Theory. Please, 'Think Trillion.'

UNIVERSE TRILLION CLUB

Each day, more new members add their name to the *Universe Trillion Club.* These people support my initiatives and endeavors to more closely examine our universe.

You can play a role.

To sign up, simply email to:

trillionist2@gmail.com

Add your name to the website list at:

www.trilllionist.com

Thanks once again for reading this theory.

Cheers to you from author Ed Lukowich.

The End

Made in the USA
Charleston, SC
29 July 2015